Nicholas Clee is a journalist and author. He lives in north London with his wife and two daughters.

Also by Nicholas Clee

Eclipse: The story of the rogue, the madam and
the horse that changed racing.

DON'T SWEAT THE AUBERGINE

THE AUBERGINE

What works in the kitchen and why

Nicholas Clee

BLACK SWAN

DON'T SWEAT THE AUBERGINE
A BLACK SWAN BOOK: 9780552778008

TRANSWORLD PUBLISHERS
61–63 Uxbridge Road, London W5 5SA
A Random House Group Company
www.transworldbooks.co.uk

First published in Great Britain
in 2005 by Short Books

This revised and updated Black Swan edition published 2012

Addresses for Random House Group Ltd companies outside the UK
can be found at: www.randomhouse.co.uk
The Random House Group Ltd Reg. No. 954009

The Random House Group Limited supports The Forest Stewardship Council
(FSC®), the leading international forest-certification organization. Our books
carrying the FSC label are printed on FSC®-certified paper. FSC is the only
forest-certification scheme endorsed by the leading environmental
organizations, including Greenpeace. Our paper-procurement
policy can be found at www.randomhouse.co.uk/environment

Typeset in 10/14pt Cheltenham LT by Falcon Oast Graphic Art
Printed and bound by CPI Group (UK) Ltd, Croydon, CR0 4YY.

2 4 6 8 10 9 7 5 3 1

For Nicolette, Rebecca and Laura

CONTENTS

Introduction to the second edition

It has been somewhat chastening to return to this book and read the advice that I gave, quite confidently, but no longer follow. While I'm still careful not to overcook the vegetables in a chicken stock, for example, I happily leave the meat and bones cooking for quite some time – and in a covered pot, too. I implied in the first edition that you could cook a joint of meat more slowly in the oven if you left it uncovered, but have since realized that the rule does not work for me, because I cannot set my oven dial lower than 130°C. I no longer add salt and vinegar to the water when I poach eggs. The blindingly obvious reason why I always overcooked rice when I pre-soaked it has at last dawned on me.

I omitted pastry and cakes from the first edition. The reason, I told myself, was that they did not suit the spirit of the book, which aimed to provide general principles around which you could improvise. With pastry and cakes, you follow the instructions, full stop. But I was kidding myself. The real reason for leaving them out was that I was useless at making them. It's a matter of dexterity, and perhaps of patience. I cannot do anything neatly. My pastry and cakes are uneven, raggedy affairs.

They do not taste too bad these days, however. They are, as I have come to admit to myself, just like any other branch of cooking: if you have a vague understanding about what the ingredients do, you have a reasonable chance of getting them to behave in the way you want.

INTRODUCTION

The most stressful event in your life, I have read, is moving house. Giving a speech is up there too. Weddings can cause a good degree of anxiety. Here's another scenario to get you in a sweat: there are eight people coming to dinner, you're attempting a dish you haven't prepared before, and it's not going well.

Why did you buy a whole leg of lamb? Boning it might have taken Gordon Ramsay five minutes, but took you forty-five. You have spent a further half an hour inserting little bits of rosemary and garlic, which stuck to your fingers, into the flesh. The cannellini beans to go with it were supposed to be cooked by now, but the one you've just tasted had the consistency of a pebble. Lacking a 'gratin dish, preferably earthenware' for the gratin dauphinois, you made do with your old Pyrex, and you stuck to the quantity of milk and cream specified in the recipe even though the liquid came only halfway up the potatoes. Now the liquid has evaporated, and the potatoes are still crunchy. Then there's the pudding: did you allow some yolk to seep into the whites when you separated the eggs? After ten minutes of beating, the whites have barely turned white, let alone risen to what the book describes as 'soft, snowy peaks'. Meanwhile your flatmate/boyfriend/girlfriend/husband/wife, having promised to lay the table and clean the glasses, is still in the bloody bathroom.

I think that we get into these flaps because food experts, and our own insecurities, have led us to think of cookery as the fulfilment of recipes. The recipe sets the standard to which we aspire. If the dish doesn't work as the cookery writers suggested it would, we must have done something wrong. We have all felt those moments of panicked helplessness when natural processes have refused to conform to our hopes.

The recipe said, 'Heat 200ml of cream and 50ml of milk, and pour over; the liquid should be level with the top of the potatoes.' But the liquid wasn't level with the top; it came only halfway up. I must have sliced the potatoes too thin, you conjecture. Or too thick. Or used the wrong kind of dish. But the recipe specified that amount of liquid, so I'd better not add any more, or something else will go wrong. Oh, no: the dish has dried up and the potatoes are uncooked. I'm a failure.

Let me illustrate the drawbacks of adhering rigidly to recipes with a couple of examples. Both dishes contain lamb, and both sound as if they should be straightforward, flavourful and unfussy.

The first appeared in my Sunday colour supplement. It's lamb cutlets with Mediterranean vegetables. You cut plum tomatoes lengthways, sprinkle with olive oil, sea salt and icing sugar, and bake for about two hours. You colour halved onions and whole shallots, sprinkle them with oil, sugar and thyme, and bake them in a foil parcel. You sauté sliced courgettes for two to three minutes; you sauté sliced aubergines for five to eight minutes; you put all the vegetables together. You wash and top some green beans; you slice fennel with a mandolin. Then you season your rack of lamb, colour it for two minutes each side in a frying pan, and put it in the oven for five minutes. Boil the beans, add them to the fennel with some olives, olive oil and basil.

Assemble the lamb, bean mixture and courgette mixture on four plates.

There is a place for this kind of recipe. That place is in the kitchen of the Michelin-starred chef who wrote it, or in the homes of enthusiastic amateurs such as those who appear on the television programme *Masterchef*. I am an everyday cook. I am never going to prepare this version of lamb cutlets with Mediterranean vegetables.

The second recipe is typical of the hearty stews you find in books celebrating country cooking:

Lard
1kg shoulder of lamb, cubed
4 carrots, sliced
3 onions, thinly sliced
1 dstsp flour
1 glass white wine
4 medium turnips, sliced
1 bouquet garni
Salt and pepper

In a casserole, heat a layer of lard, and brown the lamb in it. Add the carrots and onions and, when golden, sprinkle on the flour. Throw in the white wine, and the turnips. Add the bouquet garni, and season with salt and pepper. Simmer for two hours, adding a little water if the stew is in danger of drying out.

That looks like something I might be bothered to tackle. But I have some doubts about the details. Lamb is fatty; with lard as well, the sauce in the stew is not going to be easy to digest. Like many stews, this one has vegetables that flavour the sauce and that are also part of the finished dish. After two hours, the sauce, not the vegetables, will contain the flavour. The carrots, as anyone who has eaten overcooked specimens of this vegetable knows, will be particularly dull.

The recipe tells you to use pepper. But ground pepper, if simmered in liquid for some time, becomes acrid.

The consistency of the sauce might also be a problem. The recipe tells you to add water if the dish is in danger of drying out; but you are more likely to find that water from the vegetables and juices from the meat, with only a dessertspoon of flour to thicken them, produce more liquid than you want. Perhaps you're meant to leave the pot uncovered, to allow the sauce to reduce; but have you noticed how often recipes omit this important detail?

Lamb cutlets with Mediterranean vegetables belongs to the largest genre of recipes in books and magazines: ones so elaborate that even keen cooks will shy away from them, except perhaps on special occasions. Yet, in important respects, such recipes are not elaborate enough. They are long on instruction, short on help. Those aubergines with the rack of lamb, for example: you are going to have to fry them in batches, while they absorb a great deal of oil. Will they really cook through in five to eight minutes? I've never managed to get a sliced aubergine tender in that time. (You may wonder, too, why the recipe does not tell you, as most recipes do, to salt the aubergines in advance. We'll come to this controversial topic later.) What about the cooking time, ten minutes or so, for the lamb? How rare do you want it?

So we resort to a simple recipe, such as the stew. But it turns out to be too simple, as are many of the recipes in the book from which it comes, Keith Floyd's otherwise enticing *Floyd on France*. Again, it ignores the problems the cook is likely to encounter. A bit of extra work will produce a nicer dish. The sauce would benefit from the removal of the vegetables, and possibly of the grease too – even if you're going to eat the stew, as Floyd recommends, in the Cheltenham racecourse car park on Gold Cup day. One might add some freshly cooked, glazed turnips and carrots as garnish instead.

Here are some annoying things that authors of recipes do:

They tell you to take a piece of roasted meat out of the oven and leave it 'in a warm place'. Such as where? Your airing cupboard? A warm oven with the heat switched off might suit; but you are likely to have other food cooking in it. Unless you are the happy owner of a double oven. I am not.

They tell you to make a sauce, set it aside and keep it warm while you perform five further stages of the cooking process. See above.

They tell you to peel and slice 1kg of potatoes, and fry them in a pan 'large enough to hold them in a single layer'. That pan would be the size of a dustbin lid.

They tell you to brown onions, then to add cubed meat or mince and brown that too. You have to turn up the temperature to brown the meat, which you leave undisturbed in order for the browning reactions to take place. Meanwhile, your onions will burn.

They think that you have enough patience to wash rice until the water runs clear.

They continue to base instructions on handed-down, questionable assumptions: you must salt aubergines to 'draw out the bitter juices', fry meat to 'seal' it, or wash rice to 'get rid of the starch'.

Have you ever tried making ratatouille according to a supposedly authentic recipe? It takes for ever. You've got to cube aubergines and salt them. Then you've got to rinse and dry them. Then you fry them; even salted, they absorb a lot of oil, they stick to the pan, and they remain determinedly firm. Then you've got to chop and fry, separately, onions, peppers and courgettes. Then you assemble the ingredients with tomatoes, which you have skinned and deseeded. Life is too short to stuff a mushroom, Shirley Conran said. Stuffed mushrooms are convenience food compared to ratatouille.

The point I'm making is that we'd be better off, more relaxed, happier, better read, if we didn't follow recipes so much. What we want for the cooking we do from day to day, and even for the cooking we do when we entertain, is a set of techniques that we know will work. Then we can improvise – as we always have to do, because of the vagaries of the cooking process – without worrying that we are committing solecisms.

Julian Barnes, as well as being the author of excellent novels, has written an entertaining book called *The Pedant in the Kitchen*. His approach is the opposite of mine. He worries away at recipes, questioning how precisely to apply the instructions. I am far less diligent. I look at a recipe and think, I can't be bothered to do that. Or, that wouldn't work if I tried it. Like Barnes, though, I am curious about why writers recommend certain methods. They don't often tell you. They can be inconsistent. A writer may instruct you to simmer stock for three to four hours. You buy the writer's next book, and the stock recipe specifies a simmering time of just one hour. Why the change? I'd quite like to know.

I have come to my own conclusions about how long to simmer stock; as well as about how to make a simple stew, about whether you need to treat aubergines with salt before cooking them, and about many other techniques that are the subject of unexplained, inconsistent or baffling assertions in conventional cookbooks. This book explains why I have come to those conclusions. It includes a little science, borrowed from better-informed writers; sometimes the recommendations are a matter of personal taste, with which you might disagree.

It's also about gaining confidence in your own taste. Nigel Slater tells a story about a woman who asked him about the tablespoon of parsley he had specified in one recipe: was it, she needed to know, a heaped tablespoon or a level one? To which the polite answer is, how much do you like parsley? Or, perhaps, would you prefer tarragon? Of course, you'd be ill-advised to mess around too much with a cake recipe. But many of the choices you make in cooking are a matter of how much you like certain flavours, whether you prefer sauces to be thick or thin, and so on. I put more chilli in the pasta sauces I cook for myself than you would like, probably, or than an Italian chef would approve of, certainly. But I like chilli, so there.

What follows, then, is advice on cooking methods, as well as a collection of templates – for soups, stews, sauces and more. I have to admit, though, that I have been unable to write a book about cooking without including what you might call recipes. My advice about stock is a recipe of sorts, even though I do not give precise instructions about the quantities of meat, vegetables and water to use. I cannot describe the making of a béchamel without telling you that the proportion of 28g of butter and one tablespoon of flour (about 28g again) will make a roux of the correct consistency, and that it will thicken about 280ml of milk; once you know that, though, I hope you'll make this sauce by taking a knob of butter, adding flour and cooking the mixture until the roux feels like loose, wet sand, and then adding milk gradually until you have the consistency you want. There's no avoiding recipes when it comes to puddings, either, unless you're going to have fruit salad at the end of every meal. Maybe that's why I'm not much of a pudding maker. But I do give a few of my favourite pudding recipes here.

I'm not knocking recipes, or recipe writers. Delia Smith is, deservedly, a national treasure, and she gives instructions that produce the results she promises. All cooks need collections of recipes such as hers. Recipes give us new ideas, help us to expand our repertoires.

However, there is a telling survey suggesting that, on average, people cook no more than two dishes from each cookbook they buy. We read the recipes and think, that looks a bit tricky. Or, where am I supposed to get hold of bream? And we return to our old standbys. This book is about broadening the range of those standbys.

Weights and Measures

Metric	Imperial	Fl oz (UK)
142ml	1/4 pint	5fl oz
284ml	1/2 pint	10fl oz
568ml	1 pint	20fl oz
1.1 litres	2 pints	40fl oz
2.3 litres	4 pints	80fl oz

Metric	Imperial
28g	1oz
57g	2oz
113g	4oz
227g	8oz
454g	1lb
1kg	2lb 3oz
1.5kg	3lb 5oz
2kg	4lb 7oz

Pan and cake tin measurements

20cm = c8in
23cm = c9in
28cm = c11in

Spoons

1 tbsp = 2 dstsp = 4 tsp

1 heaped tbsp flour = c28g

Until I came to write this book, I had no idea how much Parmesan I put into a spaghetti carbonara. About 25g for two, I discovered (p118). It seems to be in excess of what most recipes recommend. Perhaps I should be more restrained: my quantity probably unbalances the dish. But I continue to grate the cheese self-indulgently, because I like strong flavours, and I worry that only a couple of tablespoons of Parmesan will fail to make much impression.

'Yes – oh dear, yes – the novel tells a story,' E.M. Forster wrote. Oh dear yes, a cookery book must have recipes. You'd be right to complain if I simply told you to use as much Parmesan in your carbonara as you liked. But the precise measurement is a necessary convention of the cookery genre. You need to interpret it – just as, Forster thought, you should look beyond the narrative surface for the most important stuff of a novel. With an outline of what works, you can decide on your own preferences. You might prefer less Parmesan. You might prefer Pecorino. Or Cheddar.

Quantities in recipes fall into three categories: guides to how much people will eat; guides to appropriate flavourings; guides to the chemical behaviour of the dish. The first two tend to be open to interpretation; the last usually has to be followed accurately. To generalize, the instructions in the savoury section of this book – with the exception of the Sauces chapter – mostly fall into the first two categories; in the Cakes and Puddings chapter, the third category is more prominent.

Imperial measurements, officially discarded, linger in the compositions of many standard recipes. You'll notice, for example, that the quantities in the béchamel sauce (p46) correspond with more memorable imperial equivalents: 1oz of butter, 1oz of flour, 1/2 pint of milk. The batter (p225): 1/2 pint of milk, 4oz of flour, one egg. Sorry: the metric versions just aren't so neat.

TEMPERATURES

Temperatures

Gas mark	Celsius	Fahrenheit
1/4	110	230
1/2	120	250
1	140	285
2	150	300
3	160	320
4	180	355
5	190	375
6	200	390
7	220	430
8	230	445
9	240	465

An oven thermometer won't cost you very much. If the reading it gives you from the central shelf of your oven corresponds with the reading on your oven dial, you have a rare cooker. A publisher friend, wanting her colleagues to try out recipes in one of the company's books, got them to test their ovens first: there was a 50°C variation between ovens at the same setting. My own oven is hotter than the dial tells me it should be.

If a recipe tells you to cook a soufflé, say, at gas mark 5/190°C, you need to be cautious if your oven will generate a heat of 220°C at that setting.

Different sources give different gas mark/celsius conversions. Some say that gas mark 4 is not 180°C, as I have asserted, but 175°C; some, that gas mark 7 should be 210°C;

and so on. I must admit that I don't know what the correct readings are. I don't trust my own oven to give them.

Put your thermometer on an oven rack and check the temperature. Now put a pan of steaming water below it. You'll find that the reading goes down. But the steam will not cause the thermometer in a 200°C oven to descend to the temperature of boiling water (100°C). In my experience, the pointer will drop to about 160°C. I cannot expect steam in the oven significantly to slow the cooking of, say, a joint of meat; indeed, it might speed up the process, because it cooks more efficiently than does radiant heat. For this reason, steam does not, as many cookery writers imply, help to keep meat moist.

Another popular way of trying to keep meat moist is by wrapping it in foil. Steam builds up inside the foil; again, the meat may emerge drier than it would have done if left uncovered.

According to Harold McGee (*McGee on Food and Cooking*), foil deflects a good deal of heat energy that radiates from the oven walls, ceiling and floor. When I put a loose tent of foil around my thermometer, left it for half an hour, and took a reading, the reading was just as high in the foil as it had been outside it; but perhaps we're talking about the kind of heat rather than the temperature. Foil does give protection to certain foods. Whole garlic cloves, for example, go dry and bitter very easily if placed naked in an oven; surrounded by foil, they become mild and tender (see p158).

Cookery books use such formulae as: 'Put in a gas mark 3/160°C oven for an hour and a half.' That's because writers don't have enough room to give, every time, instructions along these lines: 'Put in a gas mark 3/160°C oven, but keep checking to see if the liquid is bubbling too fast or too slowly; if it's bubbling too fast, and the liquid is reducing faster than the potatoes are cooking, turn down the heat; if there's little activity in the dish, turn up the heat; you want a very gentle simmer to allow the potatoes to cook while the liquid slowly reduces and thickens' (see Gratin dauphinois, p188); and, even if they did have the room, they think, with justification, that readers would find such qualifications offputting. Just give us an instruction that will work!

I am sorry to tell you that there are very few dishes that you can just put in the oven and forget about, confident that they will take care of themselves in the time the recipe specifies. Please bear this warning in mind when you read the recipes in this book, or in any other books. My oven may be hotter or cooler than yours. My roasting pan or casserole or gratin dish may be thicker or wider, or made of a different material. The meat in my stew may take longer to tenderize than the meat your butcher sold you. These factors mean that oven settings and times that work for me may not work when you try them.

It's worse than that: settings and timings that work for me one week let me down the next. I put a gratin dauphinois into the oven at 190°C, leave it for an hour, and come back to find it still swimming in milky liquid. Next time, the same settings and utensils produce a dried-out concoction sitting underneath a burned crust. I have to turn down a stew to the lowest oven setting because it's bubbling too vigorously; the next stew, in a 150°C oven, declines to show any activity at all.

You haven't failed if your food doesn't behave in the way the recipe writer tells you it should. Everyone has to improvise. The oven setting is not the point; the point is the result you want to achieve. Don't approach roasting with the notion that you have to start cooking your meat in a 200°C oven for 25 minutes; think about the browning reactions you want to generate. If 25 minutes in a 200°C oven achieves them for you, good. Don't take it as a rule that you should cook a stew in a 140°C oven; the rule is that you want the stew just to blip a little as it simmers. Turn the oven up or down until you get the effect you want. Even then, you need to be watchful. A stew that blips gently after 30 minutes may be boiling furiously an hour later.

Then, to add another layer of complication, there are fan ovens. I don't have one, so I'll quote Jenny Webb, who, as author of *The Fan Oven Book*, should be authoritative. Fan oven owners can set their ovens at 10 to 20 degrees Celsius lower than recipes advise, she told readers in a *Guardian* piece on the subject, and can expect cooking times to be reduced by 10 minutes in the hour. My advice (p217) about roasting belly pork on the floor of an oven is redundant if your oven has a fan, which in theory will make the bottom of the oven just as hot as the top.

On the hob

Here, too, you have to do a certain amount of fussing. Some foods need fast-boiling or steaming: pasta, rice, green vegetables (see the relevant chapters). Meat that you're cooking quickly (a steak, for example), or that you're browning before stewing, also needs cooking at a high heat; but be watchful. A frying pan with oil can get hotter and hotter, and the heat under it needs to be regulated if the food is not to end up blackened rather than browned.

Stews (see p227) should cook at a very gentle simmer, with only a few bubbles blipping at the surface: subjected to a pummelling from fiercely boiling liquid, meat dries up in protest. This low heat can be difficult to maintain on a hob, especially in a casserole with the lid on, because the build-up of steam will raise the pressure in the dish, causing the liquid to bubble more vigorously and at a higher boiling point. A heat disperser will help, as will leaving off the lid. But the liquid in an open pan will evaporate.

Sometimes, you want evaporation: to thicken a sauce, or to concentrate its flavour. You have to adjust the heat under a pan in order to get the rate of evaporation you want; and remember, if you want less or thicker liquid in a stew, and if the gentle simmer is evaporating the liquid too slowly for your liking, you'll have to take out the meat before you turn up the flame. If the meat endures rapid boiling, it will go tough.

In preparing recipes that involve sauces, the trick is to have the sauce at just the right consistency at the moment you want to serve it. Good luck. If you get it wrong (if, say, your tomato sauce is as thick as you want it before you've started cooking the pasta), you cannot put the lid on the pan to arrest the evaporation, because the sauce will start bubbling more vigorously and will most likely stick to the pan base. You'll just have to turn it off and warm it up again later; or, if the pasta will be ready quite soon, put the pan of sauce (or the sauce decanted into another container), covered, into the bottom of a very low oven.

EQUIPMENT

This is not a comprehensive list. I don't have very much to say about cheese graters, lemon squeezers or sharp knives, except that if you want to grate cheese, squeeze lemons or cut things up, they are, you know, just the job.

Heat disperser

A metal disc – they used to be made of asbestos – for inserting between a pan and the ring on a hob; it tempers the heat, and spreads it more evenly. Cost: £5 each, or less. Buy a couple.

Of course, it's nice to have decent, thick-bottomed saucepans. But for many jobs – boiling or steaming vegetables, boiling pasta – rubbish pans will do. If you need the pans for more delicate tasks, such as warming through mashed potatoes or other cooked vegetables, use a disperser to imitate the task – of imparting heat evenly – that a thick bottom would perform. Making a flour-thickened sauce such as béchamel in a cheap pan can be disastrous, because the pan's base gets too hot, rapidly thickening the liquid in contact with it, and causing that liquid to stick and scorch. With the aid of a heat disperser, the pan imitates a far more expensive item.

Hobs on cookers usually generate more heat, even in solid, expensive pans and casserole dishes, than the simmering of sauces, stocks and stews requires. Use a heat disperser.

Steaming basket

I have a cheap one. It's a kind of perforated trivet, with flaps at the side that enclose the contents when you put it inside a saucepan. The lids of my pans are not efficient: steam pours out of them. Nevertheless, the arrangement works fine: food in my cheap steaming basket inside my inadequately sealed saucepans cooks at least as quickly as it would if immersed in boiling water.

Steaming preserves more nutrients than does boiling. I use my steamer – controversially, some might say – for green vegetables (see p138) and for fish (p286).

Roasting pan

You need an expensive one, I'm afraid. Cheap pans will probably do for roast potatoes and other vegetables; but if you roast meat in one, you'll find that the juices from the meat will hit the pan, boil, and burn. These juices should have formed the base of a delicious gravy.

Casserole dish

Again, the price of a good one, such as a Le Creuset, is worth paying. Stews, pot roasts and the like need slow, even cooking, which the thick walls and base of a Le Creuset foster. Also, a good casserole will have a tight-fitting lid, retaining moisture. A cold stew in a heavy casserole will take a long time to get to simmering point in the oven; don't worry about that (see p236).

Cast-iron frying pan

I have one that I bought for £5. I wish I had several; then, I could use one for omelettes, and the other for heavy-duty frying of meat. As it is, my all-purpose pan makes good omelettes anyway. You 'season' the pan by pouring in some oil and leaving it on a low heat for a couple of hours. Then you must remember never to wash it in washing-up liquid, which removes the oily patina; use water, or paper towels, only. The pan develops a slick, non-stick surface, on which omelettes slide like skis on ice. Smearing the pan with a little oil before you put it away is not a bad idea. Stacking other pans on top of it may cause rusting.

A good many recipes in this book call for 'deglazing': the addition of a liquid such as wine, vinegar or water to a frying or roasting pan to dislodge the remnants of frying and incorporate them in a sauce. An acid liquid such as wine or vinegar will undermine your carefully cultivated oily patina, however. Smearing the pan with oil afterwards may be a sufficient restoration job.

The drawback of cast iron is the weight – particularly when cooking omelettes, which require a certain amount of manipulation. If you buy a lighter omelette pan, season it before you cook with it the first time, wipe it with paper towels rather than washing it, and use it for omelettes only.

Non-stick saucepan

No matter how well seasoned your cast-iron pan, you cannot use it for making scrambled eggs: the eggs will stick, and will not be enjoyable to wash off. They will stick to metal saucepans too. For scrambled eggs, a non-stick pan is best. If you have one, you might as well cook béchamel in it as well.

Ridged grill pan

Also known as a griddle. Char-grilling may not be as fashionable as it once was, but dark griddle stripes on food remain highly desirable.

A grill pan is not ideal for cooking thick pieces of food, the outsides of which will be blackened by the time the centres are hot. I use my pan for lamb cutlets or steaks; for boned chicken legs; for chicken breasts, which I slice into two or three pieces; and for bacon. I may sear a thicker piece of meat on the pan, but cook it through in the oven.

Griddling vegetables such as courgettes, asparagus and aubergines is popular; but it leaves courgettes and asparagus too crunchy, I find, and doesn't produce in aubergines the melting consistency I like.

Our enthusiasm for char-grilling, incidentally, may not be good for us. Charred food has been found to contain carcinogens, as has the woodsmoke from outdoor barbecues.

Stockpot

Your largest saucepan probably won't be big enough for stock-making. Invest in a large stockpot, which can have a second function as a pasta pan.

Food mill

Moulinex makes one called a Mouli-légumes. You turn a handle to push food through a metal disc with serrated holes; the mill usually comes with two discs, one with very small holes, one with slightly larger ones. I use it for mashed potato, pushing the potatoes down on to the disc and working the turning device over them; for soup; and for tomato sauce made with whole, chopped-up tomatoes. The mill leaves the unwanted tomato skin on the disc.

I prefer the taste – perhaps I mean the consistency – of food that has been worked manually rather than blitzed in an electric machine. Putting potatoes into a food processor, for example, is disastrous (see p181). Soup, I think, is more interesting after manual pulping than after electric blending.

Here's what else I don't use food processors or blenders for:

Mayonnaise: food-processed mayonnaise does not have the liveliness of flavour of one turned or whisked by hand.

Chopping and slicing: the chopping and slicing blades of food processors do much more damage to the cells of vegetables than do carefully wielded knives; you'll find that electrically sliced onions and potatoes, in particular, have moist and slightly slimy surfaces.

Pestle and mortar

Don't use a garlic crusher. It produces an unpleasantly strong, bitter purée. Instead, cut the garlic into manageable pieces, and grind it in a mortar with a little salt. Or you could crush it on a chopping board with the blade of a heavy knife. The salt helps turn it into a creamy pulp.

A mortar is the traditional receptacle in which to turn – you move the pestle constantly in the same direction – aioli, or garlic mayonnaise (see p42). I use it for ordinary mayonnaise, too; mine is large enough to contain the mayonnaise that one egg will produce, and because it's heavy it does not wobble about as I turn the pestle. You may find that a whisk involves less hard work.

A pestle can also, as the derivation of the word implies, make pesto (p39). Or you can use the pestle to grind spices – but remember that spices such as cumin will leave their flavour behind, to invade subsequent preparations. You might want to invest instead in the following item.

Herb mill, or coffee grinder

These are electrical devices, of course. Don't use them for chopping herbs, apart from rosemary or curly parsley: they will turn fragile leaves such as basil or tarragon or flat-leaf parsley into flavourless mush. But they are invaluable for dried chillies – have you ever tried to 'crumble', as the recipes instruct, a dried chilli? They will also grind whole spices such as cumin and coriander seeds. You may need two mills or grinders: one for spices and chillies, the other for such ingredients as parsley, breadcrumbs and pine kernels.

Balloon whisk

Beating egg whites or cream by hand is quite hard work. But by the time you've got out your electric whisk, plugged it in, whisked your eggs or cream, dismantled the machine, washed it and put it back in the cupboard, you'll certainly have spent more time on the job than if you'd done it manually. Also, hand-beating enables you to feel how your ingredients are progressing: whether your egg whites are approaching the 'soft peak' stage, for example. It's too tempting, with an electric whisk, to leave the machine running for just an extra minute, to make sure that the foam is perfect. It will be perfect one moment, and in a state of collapse the next.

Salad whizzer

A plastic bowl with a meshed plastic container inside; you put on the lid and turn the handle to spin the container, the contents of which throw off their water. Dry salad leaves will accept an oil-based dressing; wet ones will repel it.

Purists tell you that this process can damage delicate leaves. Gordon Ramsay advises that you wrap them in a clean dishcloth, hold the ends, and give it a twirl. When I tried this technique, I sprayed bits of salad all over the kitchen. So I rely on my whizzer.

INGREDIENTS

Again, this is far from a comprehensive list. The following ingredients are so common that it's worth being familiar with their traits that recipes usually lack room to explain.

Salt

I hope you aren't too maddened by my vagueness in the following pages about the quantities of salt you should use. We all have different tastes; and, if we've ever sprinkled salt on to a pile of chips, we can probably make a rough guess about how much salt we should add to the contents of a pot. I trust that you won't pour a tablespoon of salt into a stew for six people. Be cautious at first; taste as you go along, and add more seasoning if necessary.

Water evaporates, but salt doesn't. As liquid boils away in an uncovered pan, the proportion of salt in it goes up. The teaspoon of salt that may have seemed a moderate addition to a stew will become excessive if you end up with a reduced, concentrated sauce.

Salt, sprinkled on to something moist, will suck up the water. Many cooks like to sweat vegetables with a high water content such as aubergines, courgettes and cucumbers, sprinkling salt on them and leaving them to drain in a colander, for various reasons: to reduce their sponginess, so that they don't absorb so much oil when frying (aubergines); to prevent their throwing off a lot of liquid in the pan, so that they fry rather than stew (courgettes); and to concentrate their flavour (cucumbers). Sometimes this operation is worthwhile; sometimes, in my view, not (see the Vegetables chapter, p137). A second reason for sweating aubergines – one still advanced from time to time – is that the salt draws out their bitter juices (see p140).

The water-absorbing quality of salt is responsible for the widely disseminated advice that salting meat before frying or grilling will cause it to lose its juices. But the effect of the salt is negligible by comparison with that of the high heat of the pan or grill, which drives liquid from the meat at a tremendous rate. That's why a steak sizzles in the pan; that's why it shrinks. If you've salted it, at least it will have a flavoursome crust.

A soaking in brine, a solution of salt in water, will increase the tenderness and moistness of meat when it is cooked. Pork belly, brined for three days and then gently roasted, is meltingly soft and juicy (see p218).

Salt in cooking water hastens the softening of vegetables and reduces the loss of nutrients. You have to be careful: over-cooked members of the cabbage family become rank, and green legumes go grey and flabby if they steam or poach for too long.

Every cookery writer recommends sea salt (or rock salt) in preference to table salt. However, there have been tests in which experts have failed to distinguish between them. I think that I favour Maldon sea salt as a seasoning, even though I'm not confident that I could back up that recommendation by passing a salt-tasting test; I use table salt in cooking water – for pasta, for example.

Pepper

I have been even vaguer in my advice about pepper than I have about salt. It's up to you. Add it at the table, and you get the fresh pungency of the freshly ground grains; cooking gives a milder, integrated pepperiness. My one recommendation is not to add ground pepper to stocks or stews. It turns acrid after long simmering.

Oil

On cooking, extra virgin olive oil loses the fragrance and fruitiness that make it stand out from the ordinary, refined stuff. But it's so cheap these days – can it be genuine? – that I use it as an all-purpose oil anyway. In the text, I've written simply 'olive oil'. If I had the money, I'd buy a standard brand for cooking and a fancy, estate-bottled one for salads. Sad to say, I don't.

I use olive oil for salads, for pasta sauces, and for any cooking that involves Mediterranean vegetables. For other purposes, I fry with sunflower oil. Vegetable oil is fine too. Cookery writers often recommend groundnut oil, which they reckon to be the oil that least adulterates the flavour of food; but it's hard to find round our way.

One of the more baffling of the instructions commonly appearing in cookery books is that you should heat oil, for searing meat or for stir-frying vegetables, until it is smoking. If it's smoking, it's degrading, and will have a degraded flavour. It may also be harmful: burnt materials are thought to be carcinogens.

For frying at high temperatures, use vegetable, sunflower or groundnut oils, which have higher smoke points than does olive oil. Heat the pan first, so that the oil has only brief contact with it before you add the ingredients; otherwise, the oil is more likely to burn. When you can feel, by lowering your palm towards the pan surface, that the pan is hot, add the oil, swirl it around quickly, and add your meat or vegetables. Keep them at a lively, but not ferocious, sizzle.

Sesame oil has a low smoke point. Don't fry with it, unless you're going to do so very gently; instead, add it to stir-fries at the end of cooking, as a flavouring.

Wine and other alcohol

An enduring memory from *The Galloping Gourmet*, a hit TV show of the 1960s. Graham Kerr, the GG, is standing before a stewpot, brandishing a bottle. We need a little wine at this stage, he confides; then he starts pouring the bottle into the stew and, leeringly, doesn't stop. Heady days.

It was a time when the notion of alcohol in food seemed daringly sophisticated. Now, when it causes less of a frisson, we're inclined to use alcohol more sparingly, in the knowledge that too much of it can throw the balance of flavours in a dish out of whack.

Wine or other alcoholic drinks in cooked food taste overpoweringly raw and acidic unless they are reduced. Don't add them to a dish just before serving. Typically, you use the drink to 'deglaze' – loosen the residues, in order to incorporate them in a sauce – a pan in which you've fried meat and/or vegetables; you let the liquid bubble for a minute or two before adding it, with other liquid, to a stewpot (see p55), or you use it as the sauce for a sauté (p265). As it cooks, a good deal – though not all – of the alcohol evaporates; so do the volatile acids. The liquid becomes milder, and sweeter.

Vinegar may be used in the same way, and also loses its rasping qualities after simmering and evaporation.

(Other acidic foods, such as tomatoes and onions, become sweeter as they cook as well. That's one reason why you simmer a tomato sauce until it thickens, and why you soften onions in butter or oil.)

Our idea of a professional kitchen is likely to include chefs wielding spectacularly flaming pans. Setting light to brandy or other spirits is standard practice. But is there any point to it, other than to create an impressive display? My own experiments suggest that boiling is just as efficient in getting rid of the spirity flavour – indeed, that it produces a better-tasting liquid.

Stock cubes

Marco Pierre White has not, on the whole, enhanced his reputation by advertising stock cubes. Artificial versions of something you can make with proper ingredients are deeply unfashionable. But, while I am all for preparing one's own stock (see the Stocks chapter, p51), I do not have supplies of the stuff in my kitchen at all times. I think that a cube, used cautiously, can give umami to a dish – a grounding of savoury flavour. I always cut it in half, though.

I cannot guarantee that the advice in this chapter will produce flawless results every time. Some sauces are not foolproof, or even expert-proof. But I may be able to help improve your chances of success. And, if things do go wrong, at least you'll have some idea why.

VINAIGRETTE

HOW TO MAKE IT

For a green salad for 4
Pinch of salt
1 dstsp white wine or red wine vinegar
3 dstsp extra virgin olive oil

In your salad bowl, combine the salt and the vinegar; the salt will dissolve. Add the oil, and stir vigorously to amalgamate. [1] – SEE WHY YOU DO IT Just before serving, [2] tip in the salad ingredients, which should be dry, [3] and toss thoroughly.

VARIATIONS

Proportion of ingredients. The three parts oil to one part vinegar formula works well with most salad greens, imparting just the right amount of tartness. You may prefer a five-to-one ratio for bitter greens such as endives. The starchiness of potatoes can be offset with a greater proportion of acidity: I use two parts oil to one part vinegar in potato salads.

Oils. You can also use sunflower oil, vegetable oil or groundnut oil – or a combination. Rapeseed oil has enthusiasts, too.

Vinegar. You could use balsamic vinegar – although I'm not so keen on introducing a dark brown ingredient to a green salad. Or lemon juice.

Garlic. A little goes a long way in a vinaigrette. Some people simply rub the bowl with the cut side of a clove. I take a sliver, crush it with a little salt, and add it to the vinegar. Or, making a potato salad, I might simmer a garlic clove with the potatoes, slip it from its skin, and crush it with the salt into the vinegar. It becomes milder on poaching.

Mustard. I'd use 1/3 tsp Dijon mustard with the ingredients above. Like salt, it combines happily with vinegar, so you should stir it into the vinegar in the bowl before adding the oil. If you try adding it at the end, you'll have trouble amalgamating it. (But if you want grains of salt in your salad, grind them over at the end.)

Sweetness. Try adding 1/4 tsp caster sugar or honey to the vinegar.

Anything else. There are all sorts of ingredients you could add. Pepper, obviously. Roquefort or other blue cheese (a dessertspoon, say), crumbled into the vinegar and amalgamated before you add the oil. A dessertspoon of double cream to substitute for the same quantity of oil. A shake of soy sauce. Or Worcester sauce. Make a Thai-flavoured vinaigrette with a small piece of finely chopped lemon grass, finely chopped garlic, finely chopped ginger, rice wine vinegar, sesame oil, a shake of fish sauce, and chopped coriander leaves. Add herbs to the salad: chives, basil, tarragon, marjoram, parsley.

My preferred vinaigrette includes – in the proportions given above – vinegar, salt, pepper, mustard, honey, garlic (sometimes), and two parts sunflower oil to one part olive oil.

WHY YOU DO IT

1 • Amalgamation. Vinaigrette is an emulsion in which vinegar is dispersed in oil. Amalgamating the two is a much more foolproof process than is creating a mayonnaise emulsion (see p41); but the vinegar will fall away from the oil almost immediately. Give the sauce another stir before tossing the salad in it. Or make it in a jar, and shake it

2 • Late dressing. Oily dressings soon discolour salad leaves and turn them soggy.

3 • Dry leaves. Water on salad leaves will repel the oil in the vinaigrette.

PESTO

Once you've made your own, you probably won't want to go back to the bottled stuff. Bottled pesto has a stale taste, and isn't reminiscent of basil, or pine nuts, or Parmesan, or olive oil, at all. However, I do sometimes use it, in a very modest quantity (no more than a teaspoon for four people), in potato salads made with mayonnaise, or in cold chicken salads with mayonnaise and yoghurt.

HOW TO MAKE IT

30g pine nuts
1 garlic clove
Salt
Large handful of fresh basil leaves
3 tbsp grated Parmesan
120ml olive oil

Put the pine nuts in a dry saucepan above a gentle heat until they give a lightly toasted smell. Careful: they burn easily. Grind them: I do it in an electric herb mill, because bashing them up with a pestle is hard work.

Chop the garlic, then put it into a mortar with a little salt, and grind to a paste with the pestle. (Or use a heavy bowl and a wooden spoon.) Wash the basil leaves and pat them dry with paper towels. Add them to the mortar, with the pine nuts, and grind them against the bottom and sides, until you have a green slush. Add the grated cheese and blend it in. Add the olive oil, stirring it in a little at a time, until you have a thick sauce. You may not need all 120ml.

Or: throw pine nuts, basil and garlic into a food processor. Blend. Throw in the cheese. Blend briefly. Scrape this mixture into a bowl. Blend in the oil as above. (Olive oil whizzed in an electric processor seems to lose its fruitiness – a factor to consider, too, when making mayonnaise [p41] and hummus [p168].)

Check for saltiness. Add pepper if you like.

You can use pesto to dress pasta. Or dollop it into a vegetable soup. I like to stir it into boiled new potatoes. Or try this. Cut an aubergine in two horizontally, and then into vertical slices. Brush with olive oil, season with salt and bake at gas mark 6/200°C for 25 to 30 minutes, or until soft. Spread pesto on the slices, and put mozzarella cheese on top of that. Put back into the oven for 5 minutes. Eat hot, lukewarm or cold.

VARIATIONS

You can also make pesto with parsley or mint.

MAYONNAISE

'This classic sauce,' Raymond Blanc claims (in *Cooking for Friends*), 'is very simple to make.' That is an example of the Annoying Things Chefs Write. Vinaigrette is simple: it cannot go wrong. Mayonnaise, as every cook of experience knows, can go wrong: unless you're careful, it will split.

HOW TO MAKE IT

As an accompaniment to a meal for 3 to 4 people
1 egg, at room temperature[1] – SEE WHY YOU DO IT
1/2 tsp Dijon mustard
Pinch of salt
150ml oil[2]
1 tsp white wine vinegar

Separate the egg, and put the yolk into a bowl; discard the white, or save it for a mousse or a soufflé. I use a mortar, which is the heaviest and most stable bowl I own. I have in the past turned – I think that is the correct verb – the mayonnaise with a pestle, but now usually employ a small whisk. The receptacle should be stable enough to take your turning or whisking, because you cannot hold it with your free hand – you'll need that for the jug of oil. A cushion of a dry dishcloth or a paper towel on a chopping board should hold it.

Mix the yolk, mustard and salt.[3] Put the oil into a measuring jug. Pour a drop – no more – on to the egg,[4] and start turning or whisking. When you have incorporated the oil, add another drop, whisking all the while; without letting up your whisking for a moment, continue to add drops of oil, each time waiting for egg and oil to blend before adding the next.

By the time you've added half the oil, the mixture may be very thick; thin it with the vinegar, before adding more oil as before. You should be able now to add a little more oil each time, or to pour it in a slow but steady stream – but you cannot stop whisking.

Judge for yourself when the mayonnaise has the consistency and flavour you want. You may not need all the oil. If you're not eating the mayonnaise immediately, cover the bowl with cling film and put it in the fridge, where it will keep for three days. But it's best on the day.

VARIATIONS

Thin the mayonnaise with lemon juice rather than vinegar.

Leave out the mustard if you like.

Aioli is a garlic mayonnaise that traditionally accompanies salt cod, or perhaps poached chicken. It contains a preposterous quantity of garlic – six or more cloves with the ingredients given above. Chop the garlic finely, put it in a mortar with some salt, and grind it with the pestle until it turns to a creamy pulp. Or grind it on a chopping board with the blade of a heavy knife. Then add the egg, but not the mustard, and proceed as above.

You may prefer to save the tasting of such a powerful concoction until you find yourself on a sun-drenched terrace in Provence. Use only the garlic you want – but you probably shouldn't give the name aioli to the garlic-lite version.

For a **tartare sauce**, to accompany fish and chips, add to the mayonnaise a teaspoon each of chopped gherkins, capers and shallots.

Rouille is a fiery mayonnaise with harissa (see pp134 and 302).

WHY YOU DO IT

1 • Don't start cold. Mayonnaise is an emulsion, in which droplets of oil are dispersed – thanks to your turning or whisking – in the liquid of the egg yolk, and in the vinegar or lemon juice. The emulsifiers in the yolks need waking up – they're sluggish when cold, and when brought into contact with other cold ingredients. However, I usually find that I get away with using mustard from the fridge. I bring the egg to room temperature in lukewarm water.

2 • What oil? I use 1/3 extra virgin olive oil, and 2/3 sunflower oil. Not everyone likes mayonnaise made entirely with olive oil, finding it too heavy; if you do like it, be aware that, according to Harold McGee, an extra virgin mayonnaise is likely to separate an hour or two after preparation. Ground-nut oil is a possible substitute for the sunflower, as is rapeseed – although I am not terribly keen on the latter.

3 • What to add when. Salt, stirred into the yolks initially, makes them more viscous, aiding the separation of the oil into little droplets. The mustard aids emulsification too. Some recipes tell you to add the vinegar at this stage, but it's easier to incorporate the oil without it. However, you do need to thin the mayonnaise with vinegar or lemon juice once it becomes stiff, because the stiffness indicates that there is not enough liquid left to hold more oil.

4 • Drop by drop. At first, you can emulsify only a tiny drop of oil. If you add too much, you'll end up with some that has not been emulsified; then you'll add more to it, and that won't emulsify either. You'll produce either a split liquid with oil and water separate, or a sauce in which the water is incorporated in the oil rather than the other way round.

HOLLANDAISE

Hollandaise is scarier than mayonnaise, because you have to worry about heat as well. If the eggs get too hot, they will curdle. You need a double boiler, or a bowl that you can rest inside a pan so that the base of the bowl will not touch the water simmering below. Steam rising from the water heats the eggs gently.

You can try making hollandaise in a saucepan on a direct flame; but it is a lot harder to manage this feat successfully than it is to warm custard in a pan (see p320). In custard, the eggs are dispersed in milk and cream. With only eggs, butter and a little vinegar or lemon juice in your pan, you'll have to work very carefully to stop the eggs catching on the base and curdling.

This sauce goes wonderfully well with asparagus, and

with poached fish. It's not always practical: as you'll see, it requires undivided attention, which you may not be able to spare if you have lots of other things on the go.

HOW TO MAKE IT

Enough for 4 people
3 tbsp vinegar
2 egg yolks
150g butter, cut into six pieces

In a saucepan on the hob, boil the vinegar until it has reduced by a third. Bring water in a saucepan to a simmer, and turn down the heat so that the water merely shows bubbles rising to the surface. Place a bowl inside the pan, making sure that the base of the bowl does not touch the water. Put in the egg yolks and vinegar, [1 – SEE WHY YOU DO IT] and start whisking. Add a piece of butter, and whisk it in until it has combined with the eggs; add some more; and so on. Continue whisking until the sauce thickens; turn off the heat immediately when it does. Add salt, and pepper if you like.

Herbs go well with a hollandaise; tarragon, particularly so.

You can hold the sauce for about 20 minutes above the hot water, provided that you whisk it from time to time.

WHY YOU DO IT

1 • Helpful vinegar. It's easier to add vinegar or lemon juice to a mayonnaise after you've incorporated a certain amount of oil. But in a hollandaise, the vinegar protects the yolks against the curdling effect of the heat (cf *McGee on Food* *and Cooking*).You could also add lemon juice to the yolks.

If the sauce separates, you might be able to rescue it by adding an ice cube and beating frantically. Then warm the sauce, very carefully, again.

If the worry about curdling doesn't kill you, the cholesterol in the sauce probably will.

BÉCHAMEL

HOW TO MAKE IT

28g butter
1 tbsp (28g) plain flour
280ml milk

Melt the butter over a low heat in a (preferably non-stick) saucepan. Add flour to make a roux,[1 – SEE WHY YOU DO IT] and cook very gently, stirring, for 1 to 2 minutes; the roux should have the appearance and consistency of golden, wet sand, but should not darken. Turn up the heat a little, add a modest portion of milk,[2] and stir it in; when it has formed a thick, smooth paste, add some more milk, stirring until that, too, has been incorporated. If the liquid is bubbling away before the new milk is incorporated, take the pan off the heat and stir until the mixture turns smooth. Then return to the heat, and continue adding milk bit by bit until you have a smooth sauce of the consistency you want, and let it bubble for a minute or two;[3] if you're going to put it into a serving bowl, be aware that it will thicken considerably as it cools.

VARIATIONS

Some recipes tell you to flavour the milk first, as you would for a bread sauce (see p49). The extra flavour would be nice; but I cannot be bothered. The béchamel is usually going to be a part of a dish, rather than the main element of it.

The most common addition to this sauce is **cheese**: mornay sauce is the official title. You use it for a macaroni cheese (see p127), of course, or for eggs Florentine (p80), or perhaps to pour over cooked leeks, which you then bake to make a gratin. Half a teaspoon of mustard brings out the flavour of the cheese; a pinch of nutmeg enhances the sauce too.

Adding chopped **parsley** makes a sauce to accompany fish or gammon. I like to use fish stock (p57), or the stock made from boiling the gammon (p264), and finish the sauce with cream; or to use half stock, half milk.

You can add a beaten egg to the sauce made to cover a **moussaka** (p258) – allow the sauce to cool first, though, or the egg will curdle. One egg is right for the quantities given above – and you should make the sauce quite thick. The moussaka topping will have a deliciously light, custardy texture.

WHY YOU DO IT

1 • The roux. The butter both enriches the sauce and disperses the flour before you add the liquid. Coated in fat, the flour is less likely to clump together to produce a lumpy sauce. Cooking the roux for a short time also softens the raw flavour of the flour; that flavour is less assertive in the Italian flour known as '00'.

The derivation of 'roux' is 'red', suggesting that the flour should be cooked long enough to change colour. In a white sauce, you don't want the colour to change; but in a gravy or other brown sauce, you do – and they involve the same process as making a béchamel. You cook the flour in the fat long enough to let it caramelize, and to contribute to the colour of the sauce, before adding stock. A browned flour, however, is less effective as a thickening agent.

2 • Warm milk or cold? You can, as various recipes advise, warm the milk first; your sauce will thicken much more quickly. But, provided you amalgamate milk and sauce with care, you are not more likely to get lumps if you start with cold milk. In either case, it's easier to add the milk on the heat, so that the sauce thickens as you go, rather than away from the heat, when you'll have to thicken the whole lot at once.

3 • Simmering. You may have been told to simmer a béchamel for half an hour or an hour; the instruction sometimes comes with the gloss that simmering gets rid of the floury taste. That is not the point of the simmering; it's the point, as I've said above, of cooking the roux. Chefs simmer a béchamel in order to allow all the starch granules to dissolve, creating a very smooth texture. Home chefs need not be so fussy.

VELOUTÉ

A velouté is a roux-based sauce made with stock. See, for example, Blanquette on p252.

BREAD SAUCE

Gordon Ramsay thinks it's horrible. I'm wary of arguing with a three-star chef, and particularly with him; but I love it.

HOW TO MAKE IT

For 4 (though I might eat it all)

Bring 280ml milk slowly to the boil with half a peeled onion studded with a couple of cloves, along with a little salt, a bay leaf, a few peppercorns and a pinch of nutmeg; turn off the heat, cover, and leave to infuse for half an hour. Strain into another pan through a sieve, add breadcrumbs, and warm through.

I'm afraid I don't know how many breadcrumbs you'll need. Add some, stir and simmer; the sauce will continue to thicken, so err on the side of too few at first (unless you want an excuse to add more milk and create more sauce – but of course this milk will be unflavoured). When the sauce has the consistency you want, take it off the heat and stir in a walnut-sized knob of butter. You could add a couple of tablespoons of cream too. You can leave the sauce and warm it up later, by which time it will have become very thick and will need loosening with a little more milk.

You should not think of making stock as one of those
operations that only dedicated cooks, blessed with an
abundance of time, perform. It requires little effort, and
can often make use of ingredients you might otherwise
have thrown away. Your home-made stock will add
depth of flavour to soups, stews, risottos and many
other dishes.

CHICKEN STOCK

HOW TO MAKE IT

1kg chicken wings
Water to cover
Vegetables – some or all of the following, roughly chopped:
2 onions
1 carrot
1 leek (cleaned and minus the rough green leaves)
2 celery sticks

Put the chicken wings into a stockpot, casserole dish or
other pan with a capacity to hold all the ingredients with
room to spare. Cover with cold water by no more than 5cm.
1 – SEE WHY YOU DO IT Bring the pot to simmering point. Cover,[2]
and simmer on the lowest flame for 3 hours or longer,[3]

checking on the water level from time to time and topping it up if necessary. Throw in the vegetables, and simmer for a further 30 minutes.[4] Strain into a bowl through a colander, gently pressing down the meat and vegetables to extract as much juice as possible. Chill the stock by placing the bowl carefully in a tub of cold water (making sure the water does not rise up and pour in through the lid, obviously). Once the bowl is cool, put it in the fridge.[5]

VARIATIONS

This pale stock is known as fond blanc. To make a darker, more flavoursome variety, roast the wings first. Anoint them with a little oil and put them into a hot oven (gas mark 6/200°C for about 30 minutes, or until browned).

Your butcher may sell you cheap chicken carcasses. You could roast them first, too.

If you roast the chicken wings first, deglaze[6] the roasting pan with half a cup of water, and add this liquid to the pot.

Don't throw away the remains of a roasted chicken. Gather all the bones and leftover bits from people's plates, and put them into the stockpot with the carcass, again adding vegetables later. If you're lucky enough to have bought a chicken that came with giblets, and haven't used them already for the gravy with your roast, add them; but leave out the liver, which gives the stock a bitter taste.

You don't have to peel the onions: there's goodness in the skin. Be cautious with the carrots, which add a good deal of sweetness to the liquid. Mushrooms and unpeeled garlic cloves are other possible ingredients. You might add herbs, too: parsley, thyme and bay work well.

If you've used chicken wings, don't throw them away. They are forgiving things, and remain moist and tender even after long cooking. Allow them to cool, and cover with a marinade (see p271). Heat them in a gas mark 6/200°C oven for 15 minutes, or under a grill.

WHY YOU DO IT

1 • Don't drown the stock. You can add more water later, if you have too little. The flavour of a concentrated stock that has been diluted will be better than that of a watery stock that has been boiled down. You know that you've got the ratio of water to meat right when, after cooling, the stock has turned into a jelly (see point 3, below).

Don't add salt to a stock. You may want to reduce the liquid later; that will increase the concentration of any salt in it. Add whole peppercorns, if you like, but not ground pepper, which after long simmering will impart an acrid taste.

2 • Covering the pot. If you want to create a restaurant-standard, clear broth, don't do this. As the liquid comes to a simmer, scum created by the connective tissue in the meat and from the bones rises to the surface. You can spoon it off and throw it away. But further material of this kind will continue to rise, and, if not discarded, will sink back into the liquid and discolour it. Do you mind? If you're a home cook, and if you're not planning to show off to your friends by producing a translucent consommé, probably not. This stuff won't do you any harm. If you're a chef, you may want to cook your stock uncovered, with the merest blip showing on the surface of the liquid from time to time; and even after that you may need to clarify the liquid with egg whites. (You whip the egg whites, add them to reheated stock, and discard them once the impurities have adhered to them.)

I cover the pot, even though the build-up of steam under the lid causes quite vigorous bubbling. I do so for the sake of simplicity, knowing that I can leave the stock without worrying too much about evaporation, and also that I am ensuring the maximum possible conversion of collagen to gelatine (see the next point).

3 • Cooking time. I don't think that any other preparation is the subject of such variations of advice about cooking times. 'Eight hours with the water steaming but not bubbling will reward you with a jewel-bright, flavour-packed consommé,' Richard Whittington writes in *Home Food*. Other writers do not go quite so far, but usually agree that long cooking is good. But Michel Roux advises a one-and-a-half-hour simmer, asserting that 'long cooking can be detrimental'. Most radical is Shaun Hill, who in *How To Cook Better* advises a simmering time of 40 minutes.

Simmering the meat and bones extracts flavour and converts collagen – a protein from the bones, skin and connective tissue – into gelatine. A satisfying stock will set to a jelly in the fridge, and will have a wonderfully rich and unctuous quality. To achieve this result, you need to simmer the bones in particular for a while – I usually allow 3 to 4 hours. Collagen begins to convert to gelatine at 70°C, so you can cook the stock uncovered at a temperature below boiling point, if you like; but a slightly higher temperature, which you get in a covered pan, will convert the collagen from the bones more efficiently.

4 • Don't overcook the vegetables. The drab quality of some stocks comes, in my view, from overcooked vegetables rather than from overcooked meat. The flavour from carrots and green things seems to grow stale after a certain point – this is worth remembering when you make soups, too. Onions do not cause this problem, and won't do any harm if added to the pot at the beginning. But keeping them in the pot for longer than the time necessary to extract their flavour won't bring any benefit, either.

5 • Rapid chilling. The Food Standards Agency advises that you cool stocks within 2 hours and consume them within 2 days, or that you conserve them in the freezer (ice trays are useful containers). I must admit that I am not scrupulous in following this advice, allowing my stock to cool in the pot before straining and refrigerating it, and keeping it for up to a week. I reckon that, as long as my stock still smells fine, and even though the jelly may have collapsed a little, it probably won't do me much harm. So far, so good. But I should tell you to do as I write, not as I do.

Don't put a hot container in the fridge. It will raise the temperature, and cause condensation.

Don't reheat stock more than once, the FSA adds. There may be some bacteria that will resist the rise in temperature; and, as the stock cools again, they will multiply.

6 • Deglazing. This is a term you'll come across a lot in descriptions of the making of gravies and sauces. As meat browns, it undergoes what are known as Maillard reactions – chemical changes that produce a great deal of flavour. The flavoursome residues of those reactions stick to the bottom of pans in which you have roasted or fried meat. You can incorporate that flavour in sauces by reheating the pan and adding liquid – water, wine or other alcohol, cream; the liquid will release the residue, which will dissolve as you stir. In the case of your roasted chicken wings, remove them to the stockpot, and put the roasting dish on to a medium heat; add water, and scrape and stir with a wooden spoon until you've lifted all the residues. Pour the liquid into the stockpot. If there are still bits of chicken left in the roasting dish, repeat the process.

A note about nutritional value. Stock 'has no food value apart from some minerals', Good Housekeeping (*The New Cookery Encyclopedia*) asserts. But, as we've already seen, it has the protein gelatine, and probably some remaining collagen too. Some of the nutritional qualities of the vegetables will disperse, but others will remain.

VEAL STOCK

Good butchers will sell you veal bones. Ask for them to be chopped up. Brown them in the oven, then follow the procedures above. The gelatinization process from these bones takes longer, so you need to simmer this stock for about 3 hours.

You can also make a stock with lamb bones. It is insistently lamby, and 'can only ever really be used with lamb dishes', Alastair Little and Richard Whittington (*Keep It Simple*) advise. I make lamb braises, or gravies for lamb roasts, with chicken stock.

FISH STOCK

1 onion
1 leek
1 celery stick
Butter
Trimmings and bones from white fish[1] – SEE WHY YOU DO IT
Parsley
Water to cover

Chop the vegetables finely, and soften them for 10 to 15 minutes in butter.[2] Add the fish, parsley and water to cover – as in the chicken stock, by no more than a few centimetres. Bring to a gentle simmer, and continue to simmer for 30 minutes.[3] Strain as above. You'll probably need to use it within 24 hours.

VARIATIONS

Fennel, with its aniseed edge, is a nice flavour in fish stock. You could include wine or vinegar; add to the sautéed vegetables and let the liquid reduce by about half. The reduction will reduce the acidity, which might otherwise inhibit the cooking of the vegetables.

1 • Use white fish. Oily fish, such as mackerel and herring, will not make a digestible liquid. But you can use for a stock the shells from shellfish, bashing them up first (Fergus Henderson in *Nose to Tail Eating* suggests you might use a hammer for this job). They will be salty.

2 • Softening. For advice on this procedure, see p164. It sweetens the vegetables; a less harsh edge to their flavour is welcome in a fish stock.

3 • Half an hour is enough. Fish bones cooked for longer than half an hour may break up, adding bitterness to the liquid.

VEGETABLE STOCK

This is not, officially, a stock – it's not recognized in classical French cuisine. But it's very handy as a standby on all those occasions when you might use a meat or fish stock but, for reasons of culinary or ethical inclination, cannot. Use it in soups to give more depth of flavour than water would provide; try it in a vegetarian risotto.

There's no point in giving a recipe. Use onions, carrots, leeks and celery; fennel bulb, garlic (left whole – chopped, it will be too dominant a part of the stock) and mushrooms are optional. Add, too, the herbs recommended for the stocks above.

Cut the vegetables up small. Soften them in butter or oil for 10 to 15 minutes, then add the herbs and water to cover. You can cover the pan; the reasons given above for leaving a pan uncovered do not apply here. Simmer for no more than 30 minutes. All the flavour will be in the liquid by then; you will only dull the flavour by cooking the stock for any longer. Strain in a sieve.

Symphony orchestras cannot make their reputations playing Brahms, it is said; you might also say that chefs do not make their reputations cooking soups. Of course, there are soups that are gastronomic marvels. If you want to be gastronomically marvellous, you'll probably find a recipe from Gordon or Jamie or Nigella to suit. But the essence of soup seems more humble. Precise instructions are beside the point; this is food that you improvise, making use of ingredients to hand.

The first is a broth: stock (which might also be vegetable or fish stock) is heated, flavoured, and added to a starchy ingredient. In the second, vegetables are cooked together, sometimes after an initial softening in butter or oil; the soup is usually blended. In the third, involving dried legumes such as beans or lentils, some of the ingredients have different cooking times, and are therefore cooked apart and assembled just before serving.

For fish soup, see p300.

A VERY SIMPLE SOUP

For 4

Make a dark chicken stock (see p52). Heat 4 bowlfuls of it – or more, if you want second helpings – in a saucepan, with salt to taste. In 2 tbsp olive oil, fry 2 cloves of finely chopped garlic until it starts to turn brown; be careful, because the stage between browning and burning is brief. When it's cooked, add a whizzed, dried chilli if you like. Toast 4 pieces of white bread, and put each, torn up as necessary, into a soup bowl. Pour some oil and garlic over the toast in each bowl, and pour the hot stock over that.

VARIATIONS

When the stock is simmering, cook 350g **spinach** in it: cram the spinach into the pan, and wait for it to start to collapse; then stir it into the liquid – it should be tender in no more than a couple of minutes. Divide the stock and the spinach between the 4 bowls with their bread, garlic and oil.

Add 4 ladlefuls of stock to the garlic and oil; then pour in 2 **tins of tomatoes** with their juice. Simmer for 10 minutes or so. Pass through a food mill. Warm the soup again; pour it over the toast in the bowls.

Skin 4 **plump tomatoes** (see p195), cut them up, and add them to the browned garlic and oil; cook until the tomatoes go mushy; add a portion of stewed tomato to the bread before pouring hot stock over them.

Spread a little olive oil over each side of the **crustless bread**, and bake in the oven until brown; or grill on a ridged pan.

Pasta or noodles are an alternative starchy ingredient to the bread. Heat the stock and throw in (for 4) 150g small pasta shapes; divide between each bowl a tomato stew made as above; pour over the broth and cooked pasta.

Flavour the stock with chopped chillies (as many as you like, with the pith removed if you don't want the soup to be too fiery), a teaspoon of minced ginger, a couple of bashed-up stalks of lemon grass, and a couple of cloves of crushed garlic; add 200g to 250g cooked and drained noodles (see p129) to warm through. (Noodles will make the soup very starchy if you cook them in the stock.)

The deliciousness of these soups lies in the distinctive deliciousness of the ingredients. In leek and potato and related preparations, a bit more blending takes place.

LEEK AND POTATO SOUP

HOW TO MAKE IT

For 4

Stock or water

30g butter

1 onion, chopped

2 leeks, tough green parts removed, sliced and washed

3 medium-sized potatoes, peeled and cut up

Salt

Heat the stock or water.[1] – SEE WHY YOU DO IT Melt the butter in a saucepan and soften the onion in it (thin it with a little olive or sunflower oil if it shows signs of catching, or add a little water to the pan – see p164). Cook until the onion starts to turn golden: about 10 to 15 minutes. Add the leek, and stew until it softens: about 5 more minutes.[2] Add the potatoes, and pour in the hot stock or water to come to the level of the vegetables at the top. Add salt to taste,[3] and simmer, uncovered,[4] until the potatoes are soft: about 15 to 20 minutes. Blend the soup by pushing it through a food mill, or mash it with a potato masher.[5] For a richer finish, swirl in a little butter away from the heat (in the hope that it won't split – see p269), or some double cream (1 tbsp, say), or some sour cream.

VARIATIONS

An even simpler version of this soup has the leeks and potatoes thrown into boiling, salted water, cooked, blended and enriched with butter. I prefer the aromatic base of butter-softened onion – a base that may underlie any number of vegetable soups.

Soften an onion or two, and add whatever vegetables you like, with as much liquid as you like (I prefer thick soups). Add garlic to the onion base; I usually soften it for a minute in the butter or oil before adding the chopped onion (see p158).

Spices go with most vegetables, and in soup they complement the root vegetables particularly well, as Jane Grigson famously showed with her spiced parsnip soup (from her *Vegetable Book*). Carrot and coriander has become a ubiquitous combination: for a soup with four carrots, warm a teaspoon of coriander seeds in a dry saucepan until they give off a toasted aroma, grind them in an electric mill or with a pestle, and cook with the softened onion (and garlic) for a minute or two before adding the carrots and stock or water. A potato would give the soup a little more body. After blending, and just before serving, enrich the soup with a little cream, and add some fresh coriander leaves.

WHY YOU DO IT

1 • Add hot liquid. It's in order to keep the simmering time short. As do stocks (see p51), soups lose freshness of flavour if their ingredients are overcooked. Take the pan off the heat as soon as the vegetables are tender.

2 • Sweating the alliums. It removes their harsh notes, sweetening them. You don't have to sweat the leek, which is far less sulphurous than the onion; it will retain more of its raw character if you skip that phase. Some recipes insist that you turn the potatoes with the other vegetables, coating them in fat, before adding the liquid; I doubt if this procedure has any effect on the finished dish.

3 • Salt and pepper. Add salt when you start simmering: it will season the soup, and hasten the softening of the vegetables. But pepper can turn bitter in simmering liquid. Add it at the table, if you want.

4 • Uncovered. I don't have full confidence in this instruction, but I have gained the impression that vegetable soups from lidless pans have fresher flavours than do those from covered ones. Perhaps you're more likely to overcook covered soups; or perhaps the volatile acids, which would otherwise escape but which now condense on the lid and fall back into the liquid, are an unwelcome presence. I don't know. Your own experience may make you think that I'm talking rubbish.

5 • Blending is not compulsory – a minestrone being a famous example of an unblended soup. But the ingredients have given up their flavours to the liquid, and are themselves dull, I find. I like to blend them, but to leave the soup containing hints of its ingredients. Putting this soup into an electric blender is a bad idea, just as is mashing potato in a food processor: the blades activate a good deal of gluey starch. However, a food mill is not always ideal. It has a bit of trouble with spinach, and struggles too with legumes such as peas, lentils and beans, trapping their skins. For soups without potatoes, I often use a hand blender, though I do find that it over-processes some portions of the soup as I push it around the pan in the search for unblended bits.

PUMPKIN OR SQUASH SOUP

You could peel the squash, cut it up, and cook it as you would in the soups described above. But it's even nicer if you concentrate the flavour. Baking the chunks is one solution; but, because peeling squashes is tricky and wastes a lot of good vegetable, I prefer to cut them in half, anoint them with oil and seasonings (I like cumin seeds in addition to salt), and bake them for about 45 minutes at gas mark 6/200°C, or until tender. Then it's easy to scoop out the flesh, add it to a base of sweated onion and garlic with enough water or stock to make a thick soup, blend, and warm through. Fresh coriander and cream are nice last-minute additions.

LENTIL SOUP

A good many recipes for lentil and bean soups are dictated by the cooking time of the most resilient ingredient. In other words, you overcook the carrots while you're waiting for the dried legumes to soften. It's a mistake to think that the flavours are blending and maturing; they're getting tired, as they do in an overcooked stock (see p54, or Add hot liquid, p65). The way to get the flavours to blend and mature is not to boil the ingredients for hours, but to leave the cooked soup, as you might a stew, overnight.

For the best-tasting soup, cook each ingredient for only as long as it needs to become tender. If you have ingredients with big differentials in their optimum cooking times, cook them apart, combining them at the end.

HOW TO MAKE IT

For 4
250g lentils
2 onions, chopped
2 cloves of garlic, finely sliced
1 1/2 tbsp olive or sunflower oil
1 tsp cumin seeds
1 tsp coriander seeds
1/2 tsp turmeric

Version 1

Some varieties of lentil cook quickly: red ones take about 20 minutes, and green ones such as Puy are usually ready in half an hour. Make an onion and garlic base (see Leek and potato soup and variations, p63); for a spicy soup, add to this base the cumin and coriander, toasted and ground (see p64), and cook them for a couple of minutes. Wash the lentils, add them to the spicy onion base with a little salt and the turmeric, and cover with water or stock; cook until the lentils collapse, and blend. Check the soup from time to time, adding water if necessary. (For advice on cooking lentils, see p169.) Add chilli powder or cayenne pepper with the turmeric if you like. Don't blend the soup if you don't want to.

Version 2

Or you could keep the flavours distinct: wash the lentils, and cook them in water or stock with a little salt, 1/2 tsp turmeric, and cayenne pepper if you like; in a separate pan, fry the onions and garlic in oil until both are golden (this may take 25 minutes or more – see p164); add the toasted and ground spices and cook for a couple of minutes longer; stir this mixture into the lentils a minute before you take the soup from the heat. You could mash up or blend the lentils before adding the spicy onions, or blend the entire soup after adding them. Either way, their flavour will be more apparent in this version than in version 1.

VARIATIONS

Carrots add sweetness, but lose character if they spend too long in simmering liquid. The answer is to chop up a carrot into dice (cut a carrot into three or four pieces horizontally, cut these pieces vertically into fine slices, cut more fine vertical slices at right angles to the first ones, then cut fine horizontal slices), and add them to the onion/garlic/spice mixture, cooking until the dice are soft – that may take 20 minutes or more. Then stir into the soup, and blend or not as you wish.

Spinach is a lovely complement to spiced lentils. The most economical way to cook it would be to add it to the soup when the lentils are nearly ready, wait for it to collapse a little, then stir it into the liquid below. Unfortunately, spinach throws off a terrific quantity of water, and can turn a thick soup into a thin one. So, although the method is wasteful of whatever nutrients disappear with the water that the leaves disgorge, I cook spinach apart (see p193), drain it, squeeze out the excess liquid, and then add it to the soup.

Pep up the soup with some **lemon juice**: 1/2 lemon, added just before serving, for the quantities above. Garnish with coriander leaves or parsley. Stir in, away from the heat (to avoid splitting), a tablespoon of thick yoghurt.

BEAN SOUP

For 6

250g dried beans (kidney, haricot, cannellini, or borlotti),
    soaked overnight

2 onions: 1 whole, 1 chopped

3 garlic cloves: 2 unpeeled, one finely chopped

2 dried chillies (optional)

Olive oil

2 celery sticks, chopped

Chicken stock (optional)

1/2 savoy cabbage, cut in half again, white stalk removed,
    cut into strips, then into shreds

Drain the beans, cover by about 4cm with fresh water (I use filtered water, because my hard water tends to toughen the skins) in a saucepan, bring to the boil, and skim off the albuminous scum; add the whole onion and the two whole, unpeeled garlic cloves, and the chillies if you're using them. Simmer, with the pot partly covered, until the beans are tender. It may take one hour; it may take three, or more.[1] – SEE WHY YOU DO IT Top up the water if necessary, but don't drown the beans; you want to be left with a concentrated liquid.

Warm a layer of olive oil in a saucepan, and gently fry in it the chopped onion, celery and garlic, adding more oil if the vegetables show signs of catching on the pan; give them 15 to 20 minutes, until they turn golden and sweet.

When the beans are soft, remove half of them with a slotted spoon. Remove, too, the onion and chillies, and throw them away; but keep the garlic cloves, slipping them from their skins and adding them to the beans you have lifted from the pan. Push the beans and garlic through a food mill; or, if you want a simpler task and are happy with a coarser result, mash them with a potato masher.

If your whole beans are not sitting in more liquid than you want in your soup, return the mashed beans and garlic to them, along with the softened onion, celery and garlic. Return to a simmer. You're about to cook the cabbage in this soup; do you need more liquid? If so, and if you're not vegetarian, add some hot chicken stock.[2] Shove the cabbage into the simmering pot, turning up the heat and pushing down with a wooden spoon to get it into the liquid. Once the soup is simmering again, the cabbage should need no more than a couple of minutes to cook. Check the seasoning, and serve.

Going back a stage: if the whole beans, once cooked, are sitting in too much liquid, drain them, but reserve the liquid. Allow the beany sludge of the drained liquid to settle at the bottom, then remove the thinner liquid from the top – but keep that too. Mix the mashed beans and garlic with the drained, whole beans, and return the liquid to them – first the sludgy, tastier stuff, then the thinner stuff if you need it – until you have the consistency you want.

WHY YOU DO IT

1 • I could go on. For more – quite a lot more, if you can take it – about cooking dried beans, see p144.

2 • Hot stock. It's safer to heat the stock before adding it to the soup. Food hygienists say that reserved stock should be brought to the boil and simmered for a few minutes before use.

VARIATIONS

'Call that a bean soup?' – or words to that effect – a Tuscan would scoff. A **Tuscan ribollita** might also contain cavolo nero (the Italian dark-leafed cabbage), carrots, potatoes, tomatoes and leeks, and be served on toasted bread rubbed with garlic. The soup I describe above is much humbler; but, with a little trepidation, I would say that it has the advantage over many more authentic recipes of not containing vegetables that have been boiled for hours on end. If you want carrots in the soup, don't simmer them for longer than they take to cook (remove them from the simmering liquid if you need to). Stew tomatoes with a little olive oil, and add them in the final, warming-through phase. Add chopped leeks with the cabbage; or sauté them, adding them when the onion, celery and garlic have been in the oil for 10 to 15 minutes.

This could be a **meaty soup**. Simmer a ham hock or piece of gammon with the beans, removing it when the meat is tender. Then cut up the meat and add it to the soup at the end, to warm through. Or use a piece of bacon or pancetta: simmer the rind with the beans, for flavour and for the gelatine it imparts to the liquid; cube the meat, and brown it in oil, adding the onion, celery and garlic to the same pan to soften. A simpler version: cook the beans in the usual way, puréeing half, all or none of them as you see fit; then fry lardons and garlic, add them to the soup, and warm through.

Simpler still: cook the beans as above, with an onion and two unpeeled garlic cloves; discard the onion and the skins of the garlic, liquidize the beans and garlic pulp with enough liquid to make a creamy soup; warm through, check the seasoning, and serve with extra virgin olive oil to drizzle on the soup at the table.

If you're using **tinned beans**: they don't need cooking, of course, so they get treated as would any other cooked vegetable in a soup – added at the end, and warmed through. One possibility: simmer a ham hock or piece of gammon with a clove-studded onion. When the meat is tender, throw away the onion. Soften onion, celery and garlic, as in the bean soup above. Liquidize these vegetables with two 400g tins of beans, drained, and enough ham cooking liquor to make a thick soup. Put in a pan, warm through, and cook savoy cabbage in the soup as above. Add cut-up ham or bacon, warm through, check seasoning, and serve. If you prefer: don't blend the soup, or blend only half the beans to thicken the liquid a little.

PASTA AND CHICKPEAS

As I've described lentil and bean soups, I feel that it would be odd to leave out a version of this Italian classic.

For 4

200g chickpeas
1 onion, peeled
2 whole, unpeeled garlic cloves
2 dried chillies (optional)
2 sprigs of rosemary (optional)
Ham hock or 150g chunk of pancetta (optional)
Tomatoes from a 400g tin, drained of their juice
150g ditalini, or other small pasta shape

Soak the chickpeas overnight, drain them and cook them in fresh water (see p167) with the onion, garlic, and dried chillies. The rosemary shouldn't really be optional; but as you have to find a muslin bag or some other arrangement for tying it up (so that the sprigs don't invade the soup and get in your way when you're eating), let's say that it is. Add the rosemary and the ham hock, or the rind from the pancetta.

Remove the ham when it is tender; the chickpeas may take some time longer to cook. If you've used pancetta, cube it, and cook it separately in a little olive oil, frying it gently until it starts to get brown and crispy.

When the chickpeas are ready, they should be sitting in enough water to contain and cook the pasta as well; keep checking on the water level as the chickpeas simmer, topping it up if necessary with hot water from the kettle. Throw away the onion and chillies; lift out the garlic, remove the pulp from the skins, and return it to the soup. Chop up the drained tomatoes, and add them to the pot; when the soup is simmering again, throw in the pasta, and turn up the heat to cook it; stop when it still has plenty of bite, because it will carry on cooking in the liquid. Stir in the fried cubes of pancetta, if using, or the ham, shredded from the bone; check the seasoning, and cover the pot for a few minutes away from the heat, to allow the pasta to finish cooking and the meat to warm through. Put extra virgin olive oil and Parmesan cheese on the table, for each person to add to the soup in his or her bowl.

If you're using tinned chickpeas: add them to a mixture of stock – perhaps one made with a ham hock and vegetables (see the bean soup variation, above) – and stew made with garlic and drained tomatoes (soften a couple of finely chopped cloves of garlic in olive oil; add the drained, chopped tomatoes and cook them, with chopped dried chilli if you like, until they thicken). Then cook the pasta in this soup, and serve as above.

EGGS, CHEESE, SAVOURY TARTS

The best eggs I have ever eaten came from a Normandy farm, now in disuse, next to the place where we spend our holidays. The chickens would roam about the neighbourhood, and would lay eggs that were glorious reflections of their liberated lifestyles: vividly yellow, rich and creamy.

After eating such eggs, you cannot help but make the connection between paler, feebler battery offerings and the sad lives of the animals that produce them. Free range or, even better, organic eggs may be a good deal more expensive; but, if you calculate the cost in ratio to nutritional value, they are still among the cheapest foodstuffs on the market.

The proteins in eggs toughen, as they do in meat, if cooked for too long or with excessive vigour. When you're boiling, poaching, scrambling or frying eggs, do it gently.

There is a school of thought that eggs benefit from being kept in a cool place, or even at room temperature. However, Harold McGee (*McGee on Food and Cooking*) warns that the salmonella bacterium enjoys the warmth, and that eggs deteriorate four times faster at room temperature than they do in the fridge. The Food Standards Agency also advises refrigeration. Try to remember to take eggs out of the fridge an hour before boiling them, especially if you don't want their shells to shatter as they hit hot water.

This book includes three recipes – for mayonnaise (see

p41), for chocolate mousse (p339), and for lemon mousse (p341) – that involve raw eggs. According to the most recent (March 2004) survey by the Food Standards Agency, one in 290 boxes of eggs on sale in the UK may be contaminated with salmonella. You don't have to worry about that when you cook eggs, because salmonella is destroyed by heat; but a raw contaminated egg will poison you. So the question you have to ask yourself, when making mayonnaise or mousse, is: do you feel lucky?

Free range and organic eggs are just as likely to be contaminated as battery ones, apparently. Nevertheless, reputable UK farms carry out regular checks on their poultry, poultry houses and feed to ensure that they are salmonella-free.

BOILED EGGS

I'm sorry: you probably know how to boil an egg. I just have a few pieces of advice, if you won't feel too patronized.

Bring the water to the boil, then turn the heat on your hob right down before lowering in the egg or eggs. You're less likely to crack the shells if the water is at a gentle simmer (as you are if you've taken the eggs out of the fridge some time beforehand), and the eggs will benefit from slow cooking. Keep the heat at a level below simmering or boiling point; just let it show a few rising bubbles.

It's impossible to give reliable timings. I'd say that, cooked at this very gentle rate, a medium egg will be soft boiled, with a runny yolk, at 5 minutes; after 7 minutes, the yolk will be part-squidgy, part runny (that's how I like it); at 10–12 minutes, it will be hard-boiled, but still moist. After that, it starts to get dry and powdery. If you're hard-boiling eggs, put them into cold water for a while when they're ready; otherwise, their residual heat will carry on cooking them, and they will develop an unappetizing grey-greenish layer around their yolks.

POACHED EGGS

HOW TO MAKE THEM

Four is usually the most you can poach at once. Crack them into separate cups.[1 – SEE WHY YOU DO IT] Bring a frying pan or broad saucepan of water – you need only an egg's depth – to the boil,[2] turn down to a simmer, and gently slip the eggs into the water.[3] Maintain a low heat under the pan, as when boiling eggs (see above). Cook for 3 to 4 minutes, and lift out with a draining spoon.

I used to put them to dry on paper towels, but found it hard to prise them cleanly from the soggy paper. So now I put them for a few seconds on to a wooden board, before lifting them on to plates.

WHY YOU DO IT

1 • **Cracking it**. Cracking eggs is one of many simple kitchen skills at which I'm incompetent. I don't trust myself to do it efficiently above simmering water. This two-stage operation is far less stressful.

2 • Cooking in water. Purists turn up their noses at egg-poaching pans. A poached egg, they insist, is cooked in water; in poachers, the eggs rest in trays, and are steamed by simmering water in the pans below. I can't see, or taste, what's so unacceptable about steamed eggs; and the whites certainly come out neater.

3 • Unadulterated water. Some recipes suggest you add salt and vinegar, which help to keep the whites tender. Too tender, in my experience: under this treatment, my egg whites become raggedy, and fail to cohere round the yolks. But eggs may behave differently in the water from your taps.

EGGS FLORENTINE

HOW TO MAKE IT

For 2

Cook and drain 500g spinach (see p193); squeeze out the water through the holes in a sieve or colander with the back of a wooden spoon, and arrange the spinach, seasoned, in an ovenproof dish large enough to hold 4 poached eggs (or 2, if you prefer). Make 4 (or 2) wells in the spinach. Put the dish into a low oven to keep warm. Heat water in a pan for the eggs. While it is coming to a simmer, make about 250ml béchamel sauce (see p46). Keep the sauce warm on a heat disperser above the lowest possible flame, stirring from time to time. Poach the eggs and drain them; take the spinach out of the oven, and arrange the eggs in the wells you made. Stir 2 heaped tbsp grated cheese (Pecorino or Parmesan, perhaps; but you could use Cheddar) into the sauce (the cheese will retain its character best given the briefest possible cooking); check the seasoning of the sauce, pour it over the eggs and spinach, and sprinkle finely grated Parmesan on top. Put the dish under the grill until the surface has browned.

FRIED EGGS

Here are two possible methods:

Warm a frying pan; add enough butter to give a generous covering of the pan's surface; when it is foaming, crack the eggs over it (or crack them into cups first, and slip them in from there). Cook for a couple of minutes, until the white is set.

Nigel Slater's method: crack eggs into separate cups; heat a generous quantity (a finger's depth, he says) of good olive oil in a frying pan, until a small piece of bread sizzles energetically in it. Tilt the pan, and slip the eggs (no more than 2 in a pan) into the oil where it is deepest, so that you get hot oil to run over the yolk. Level the pan, and fry the eggs for a couple of minutes, until the white is set; lift out of the oil with a draining spoon.

If you don't like runny yolks, a fried egg is not for you. In the time it takes a yolk to set, a frying white turns to rubber.

SCRAMBLED EGGS

Again, the secret is slow cooking. That, and knowing when to stop: the egg will carry on setting when removed from the heat. If I haven't made scrambled eggs for a while, I often find that I get my timing wrong.

Crack 2 or 3 eggs for each person into a bowl. Stir with a fork to blend the yolks and the whites.[1 – SEE WHY YOU DO IT] Add salt to taste, and 1 tsp vinegar for every 4 eggs.[2] Allow 10g butter or more (I use more) for every 2 eggs. Melt half of it over a medium to low heat in a non-stick saucepan (use another kind only if you enjoy cleaning off stuck egg – I'd rather clean the Augean stables). When the butter foams, pour in the egg, and continue to cook, over a low heat and stirring constantly, until the egg is set but moist.[3] Take the pan off the heat, add the rest of the butter, stir until it's melted, and tip the scrambled eggs on to warm plates. Add pepper if you like.[4]

VARIATIONS

Any of the **herbs and vegetables** that go well with an omelette (see p84) will make a happy alliance with scrambled eggs. You must make sure that the vegetables have disgorged their water, and won't spoil the rich creaminess of the dish. Simply adding grated cheese to the beaten eggs and cooking them together is fine, or you could try a fancier option, reducing a small glass of wine to a couple of tablespoons of liquid in a saucepan, and adding it, with seasonings and herbs if you like, to a mixture of eggs and cheese (Gruyère complements eggs particularly well).

Piperade is an onion, pepper and tomato stew with a bonding of creamy egg. For 2 people: 4 eggs, lightly beaten; 1 onion, sliced; 2 red peppers, deseeded and sliced; 1 garlic clove, finely chopped; one 400g can of tomatoes, or 4 plump fresh ones; 2 tbsp olive oil. Fry the onion, pepper and garlic in the oil until soft – about 10 to 15 minutes. (Or skin the pepper first – see p172.) Now, here is one of the rare times when it's important to deseed and de-juice the tomatoes (see p194): if using canned ones, drain them and squeeze out their juice. You need a dry stew that doesn't water down the egg. Chop the tomatoes, add them to the onion and pepper mix, and cook them until their moisture has evaporated. Add the eggs, seasoned, and cook until lightly scrambled.

Or make them **spicy**: (for 2) fry a sliced onion and chopped garlic in butter or butter and oil; add a teaspoon of toasted and ground cumin (see p133), and a chopped green chilli, with the seeds and pith removed if you don't want the heat; cook for a couple of minutes longer, and add 4 beaten, seasoned eggs; cook gently, stirring, until scrambled. This dish is better with fresh chilli than with dried chilli or chilli powder, I think; and it's also nice with the addition, at the end of cooking, of some chopped coriander leaves.

WHY YOU DO IT

1 • Don't overbeat. Blend the whites and the yolks, but leave a globby texture. If you beat the eggs until they become runny, they will be tougher when cooked. This rule applies to omelettes as well.

2 • Salt and vinegar. The softening effects of salt and vinegar, which can enfeeble the white of a poached egg, are just what you want here. If you use just the small amount of vinegar I recommend, you should not be able to taste it.

3 • Warm butter, slow cooking. You get the pan and the butter hot first because you want the eggs to start cooking as soon as you pour them in. However, you have to heat them gently; cook them on a high heat, and you produce pale eggy chunks with the consistency of foam insoles. Don't wait until the eggs look ready; take them off the heat just before then, and carry on stirring. They will continue to set; if you leave them on the heat, you'll find that they move from readiness to dryness before you can react. The second helping of butter, because it melts rather than cooks, will give a more luxurious butteriness to the eggs, and it will help to arrest the cooking just at the point of perfection. You can still ruin your creamily curdled eggs by tipping them on to scorching hot plates, which will cook them some more. For an even softer result, add 1 tbsp milk or cream for every 2 eggs.

4 • Pepper. If you want pepper, add it at the end; it discolours the eggs if cooked with them.

OMELETTE

HOW TO MAKE IT

You need a seasoned frying pan (see p25), with a slick surface. A medium-sized pan (23cm) is about the right size for a 2- to

3-egg omelette. Blend the whites and yolks of the eggs lightly in a bowl, and season them with a little salt and 1/2 tsp (for every 2 eggs) of vinegar (see Scrambled eggs, p82). Warm the pan over a medium flame; add a walnut-sized knob of butter, and twist the pan so that the butter lubricates it thoroughly. The butter should foam, but not turn brown. With the flame still at medium, pour in the eggs, and give them a quick stir with a fork to distribute the heat. The edges of the omelette will set immediately; with a fork or a spatula, pull these edges from the side of the pan, tilting it to allow the runny egg to take their place.[1 – SEE WHY YOU DO IT] When the omelette is set, but the surface still moist, tilt the pan away from you and roll the omelette over towards the far edge.[2] Tip it on to a warm plate.

VARIATIONS

Herbs: parsley, chives and tarragon are among the possibilities. Add them to the beaten egg before cooking.

Cheese: blue cheese might be a little overwhelming in an omelette, but almost any other kind will be good. Grate it and sprinkle it over the omelette just before rolling, and don't use too much: you want a lightly melted filling, not a sticky wodge.

Vegetables: onions, shallots, mushrooms, peppers, courgettes, asparagus. Cook them apart (see the Vegetables chapter, p137), make sure all their liquid has evaporated, and stir them into the eggs before making the omelette. If you simplified the process by sautéing, say, mushrooms, and then pouring the eggs over them, the browning that has taken place in the pan would discolour the omelette; also, omelettes made in that way seem to be tougher.

1 • **Quick-setting**. The process, from pouring in the eggs to tilting the omelette on to the plate, should take about a minute. If you left the omelette undisturbed, the egg layer that hit the pan first would be tough before the runny egg above it had set. By pulling in the edges, you allow the runny egg to flow on to the hot surface, and to set as quickly as possible.

2 • **Slick surface**. These manipulations will not work unless you have a well-seasoned or non-stick pan: without one, you'll end up with stuck bits of eggy mess. I often do anyway, because I lack skill in rolling an omelette neatly; but the result usually tastes fine.

FRITTATA

A frittata is an Italian flat omelette, served hot or cold. It's thicker than a rolled omelette, and cooked very gently, so that it doesn't toughen up before it's set all the way through. The medium-sized pan that makes your 2- or 3-egg rolled omelette will make a 5- or 6-egg frittata.

HOW TO MAKE IT

Crack 6 eggs into a bowl, beat them gently with a fork until the whites and yolks are just blended, and stir in 3 heaped tbsp of cheese – Parmesan, say, or pecorino, or Gruyère, or Cheddar. You may not need salt, because the cheese is salty. Melt butter

– be generous, use 30 to 40g – in the frying pan, turn the heat to its lowest, and pour in the egg mixture. Cook until the base of the frittata is well set – a process that may take 10 to 15 minutes. The top will probably still be runny. Now, you could do what the professionals do: set a large plate over the pan; rapidly turn over pan and plate so that the frittata lands on the plate upside down; slide the frittata back into the pan for a few minutes, to set the other side. Or do what I do: put the pan under the grill for a minute. Cut the frittata into wedges.

Variations include all those described above for rolled omelettes. Or add bacon or pancetta: chopped slices or cubes.

SPANISH OMELETTE (TORTILLA)

HOW TO MAKE IT

1 large Spanish onion
400g potatoes, sliced or cubed (most kinds will do; new, waxy potatoes hold their shape better)
250ml olive oil
5 eggs, beaten lightly

That's a lot of olive oil. You're part-frying, part-braising the vegetables in it. Slice the onion finely; peel and slice, or cube, the potatoes, dropping them into cold water if you need to keep them waiting for a while and don't want to discolour them. Warm the olive oil in a sauté pan, or any kind of heavy pan with a lid. Add the onions and potatoes, stir and cover, cooking them above a moderate heat. After 5 minutes or so, stir again, as the onions start to collapse. Keep monitoring the pan; be careful how you stir once the potatoes soften, because they find breaking up easy to do. When the vegetables are cooked, drain them in a sieve, reserving the oil. Stir the vegetables into the beaten eggs, and add salt to taste. Take a couple of tablespoons of the oil, warm it in a frying pan, and cook the omelette as you would a frittata (see above).

CHEESE SOUFFLÉ

Some dishes are bogey dishes, while others seem to come right every time. My mother, a very good cook, was hopeless at making chips. I have a hit-and-miss record with crackling (see p215). But my record with soufflé – a dish with a scary reputation – is good.

For 2
28g butter
1 tbsp (28g) plain flour
140ml milk
100g Cheddar (or Gruyère, or similar), grated
A few scrapings of nutmeg
Ground black pepper, or cayenne (optional)
3 eggs, separated[1] – SEE HOW TO DO IT

Make a béchamel (see p46) with the butter, flour and milk. As you can see by comparing the quantity of milk with that in the standard recipe, it will be thick. Stir in the cheese, nutmeg and pepper (if using).

Beat the egg yolks. When you're sure the sauce is cool enough not to curdle them, stir them in.

Whisk the egg whites until, when you lift the whisk from the egg, it forms soft peaks.[2]

Pour the cheese mixture into the egg white, and fold it in without beating (which would drive out the air). You use a turning and lifting motion, until the mixture is amalgamated.

Lightly smear an oven dish with oil (which is a more effective non-stick agent than butter, because it does not contain solids). In my experience, the shape of the dish does not matter. Pour in the egg and cheese mixture, and bake at gas mark 5/190°C for about 30 minutes, until risen, set and browned on top.

HOW TO DO IT

1 • **Separating**. Crack the egg on the edge of a bowl, and allow the white to pour in. Gently, with your hand held over the bowl and your fingers straight, tip the yolk on to your fingers, opening them slightly to allow further white to slip through. Moving the yolk from hand to hand can encourage this process. (I got this technique from the opening sequence of a TV biopic of Elizabeth David.)

2 • **Whisking**. Try to avoid letting any trace of yolk creep into the egg white. Don't add salt or, despite what some experts recommend, lemon juice or vinegar: as mentioned several times already in this chapter, they soften egg whites.

I use a hand-held whisk, feeling that, in spite of the work involved, it enables me more accurately to judge the progress of the foam.

Use a large bowl, and tip it towards you, so that the whisk gets access to as much egg white as possible. You should stop beating when the whisk, lifted from the foam, creates peaks, which do not subside. As when making a pouring custard (see p320), it's tempting to carry on, just to make sure you've got the right consistency. Resist. Further beating causes the peak stage rapidly to be succeeded by collapse.

TWO CHEESE AND TOAST RECIPES

Laying thinly sliced cheese on to lightly toasted, lightly buttered bread, and flashing it under the grill, produces a perfectly delicious result. (Though there's a trick in setting the grill dial so the cheese melts before the uncovered sections of toast burn.) But rarebit-type recipes are nice too. The trick here is to produce a mixture that does not run off the bread. An egg helps to set it; the ratio of egg to cheese here has worked for me.

80g Cheddar or Gruyère
1/2 tsp Dijon mustard
A few splashes of Worcester sauce
Scraping of nutmeg

1 egg, beaten
Pepper (you shouldn't need salt, because of the
    salty cheese and sauce)
4 slices bread

Mash together the ingredients (except for the bread).

You don't want the toast to burn before the cheese softens and bubbles, so set the grill to medium/low. Lightly toast one side of the bread. Turn over the slices, and toast just long enough to begin crisping the bread. Remove, and spread with butter and then the cheese mixture. Return to the grill until the cheese has browned on top.

Fried cheese sandwich

The edges of sandwiches always stick to my toasted sandwich maker, which is a pain to clean. So I do the following, usually with white bread (nicer than brown in this context, I think).

Lay slices of Cheddar, Gruyère or other melting cheese in the centre of a piece of bread. Lay another piece of bread on top. Melt a knob of butter in a frying pan over a medium heat, and fry the sandwich until browned on one side, pressing it down firmly with a heavy spatula. Remove the half-cooked sandwich to a plate; melt another knob of butter in the pan (away from the heat – it might burn otherwise), and fry the other side of the sandwich, again pressing down on it with the spatula.

THE PASTRY

The first edition of this book contained no recipe for pastry. The reason for the omission was personal: I am hopeless at making it. I did not feel that I could, in good faith, offer advice on the subject.

As I explained above, when outlining an omelette-rolling technique that I myself struggle to achieve, I am not dextrous. My efforts at woodwork at school were a jumble of ill-fitting joints, and my Airfix models were encrusted with surplus glue and misapplied paint. Today, I am incapable of wrapping a present without scrunching up the paper, or of folding a shirt without leaving it in need of another go with the iron. And I cannot rub fat into flour efficiently. When I try to roll pastry, I always get it stuck to the rolling pin and to the table, and end up with an uneven, glutenous slab with holes and ragged edges.

I can still make a tart, though. A food processor does the work of my incompetent fingers (though the machine has potential disadvantages – see below), and, leaving the rolling pin in its drawer, I simply spread the dough by hand (as recommended by Elizabeth David), or grate it (Hugh Fearnley-Whittingstall).

HOW TO MAKE IT

For a 23cm tin
70g butter
140g plain flour
About 2 tbsp iced water

Cut the butter into small pieces, and put it back into the fridge for 30 minutes. You could put the flour in its bowl there too.[1] – SEE WHY YOU DO IT

Tip the flour and butter into a food processor. On a medium speed, whizz the ingredients, in short bursts, until the butter is blended and the mixture has the consistency of breadcrumbs.[2] (Or, if you prefer, carry out this process with your fingertips.)

Tip this mixture back into the chilled bowl that had held the flour. One tablespoon at a time, sprinkle over the water, lifting and blending the mixture gently until it coheres; or stir it into shape gently with a knife.[3] Put it back into the fridge, wrapped in clingfilm if you want to protect it from the odours of other foods, for another 30 minutes.[4]

A loose-bottomed tin will enable you to transfer the cooked tart to a plate. Grease it with a little olive or vegetable oil – the solids in butter can cause sticking. Spread the pastry by hand over the bottom and sides of the tin; or grate it into the tin, and smooth it out.

Prick the pastry with a fork, lay foil or greaseproof paper on top, and weigh down this covering – with baking beans, or with uncooked rice, or, as I do, with another tin of the same size. Cook the pastry 'blind' (without a filling) in a gas mark 6/200°C oven for 15 minutes; remove the weight and the foil or paper, and continue to cook until the pastry loses all tackiness. Now it is ready for your filling.[5]

WHY YOU DO IT

1 • **Cold ingredients**. The trick in pastry-making is to minimize the creation of gluten – the rubbery, tough protein that forms when molecules in starch granules bond, with the help of water. 'Shortening' – a fat such as butter or lard – coats the grains of flour, repels water, and inhibits these chains of molecules from forming. A low temperature also inhibits gluten formation. The recipe includes no salt, you'll notice: salt 'greatly strengthens the gluten network', Harold McGee warns.

2 • **Machine or hand?** I use the machine, because I tend to botch the hand-rubbing. But it has drawbacks. The vigorous beating can cause the water in the butter to hydrate the starch, creating gluten, as can the heating effect of the rapidly whirring blade. I try to minimize these results by using the motor in short bursts. Lard, if you'd like to use it, has a lower water content.

3 • **Adding the water gradually, and gently**. This is a delicate stage of the process: you're introducing an ingredient that will cause gluten to form if handled insensitively. Do not pour water through the spout of the processor while it's whirring. As the ball coheres, it will be kneaded by the blade. Kneading is fine for bread, but not for pastry.

4 • Resting. Even with your delicate handling, the dough has developed some lengthier protein molecules. During the next 30 minutes or so, they will relax.

5 • Blind baking. Margaret Costa's *Four Seasons Cookery Book* is a wonderful work, but offers bad advice in suggesting you pour your tart filling into a case of raw dough. You end up with a soggy crust. When baking blind, you prick the pastry and weigh it down because it can buckle as the water in it steams.

Fillings

QUICHE LORRAINE

200g streaky bacon, cut into strips, or 200g lardons
A little vegetable oil
1 whole egg, 3 yolks
200ml milk
200ml double cream

Put a baking sheet on to the middle shelf of a gas mark 3/160°C oven.

Over a medium heat, fry the bacon in the oil until the fat runs.

Beat the eggs (you could use one of the surplus egg whites to brush on the surface of the pastry 5 minutes before the end of cooking, to glaze it). Then beat them with the milk and cream. Tip in the bacon, which you have removed from its frying pan with a slotted spoon; pour the mixture into a pastry-lined tart tin.

Put the tin on to the baking sheet (which helps to convey the heat), and bake for 30 to 45 minutes, until the custard is set.

VARIATIONS

There are any number. The first is that you could use **cream only**, instead of cream and milk; or 200ml double cream and 200ml crème fraîche. I find a cream-only tart a bit rich, but perhaps that is because I like greedily to eat half of one of these tarts (which might be enough for four), with just a green salad as accompaniment.

An authentic quiche Lorraine, Elizabeth David tells us, does not contain **cheese**. But I cannot resist adding Gruyère, or Cheddar, or Parmesan, or some other melting cheese – about 60g.

Vegetables: cooked onions, mushrooms, spinach, leeks and asparagus are the obvious ones. You need to cook the first four so that they do not leak water into the custard (see Vegetables, p137). Sweat the onions and leeks gently, perhaps in a covered pan, until they are thoroughly wilted and glossy.

Nutmeg is a nice addition to the custard. Be careful with the salt if you're using bacon and/or cheese.

CHEESE TART

This is an adaptation of an Elizabeth David recipe.

> 28g butter
> 28g flour
> 140ml milk
> 60g Gruyère, grated
> 2 eggs, separated
> Pepper, dash of nutmeg, dash of cayenne or chilli pepper
> 1 tbsp Parmesan, grated

Make a double-thick béchamel (see p46) with the butter, flour, and milk. Stir in the Gruyère, egg yolks, and seasoning. Add a little salt, if you don't find the cheese salty enough.

Whisk the egg whites until they form peaks (see p89). Fold them into the thick béchamel; or fold the béchamel into the eggs, if the bowl is easier to work with than the pan in which you made the sauce. You lift and turn, until the egg and sauce are blended.

Spread the sauce into the pastry case. It does not appear to be a particularly generous filling; but it will expand. Sprinkle the Parmesan on top.

You can bake this at a higher temperature than you would a tart with a custard filling. Try gas mark 5/190°C, until the soufflé has lifted and is brown on top.

ALSACE ONION TART

A pastry case with a flour-based filling. The Alsatians like food that fills them up. The recipe is from *Alsace: The Complete Guide* by Vivienne Menkes-Ivry.

> Lard, or butter and oil
> 3 medium onions, sliced
> 57g butter
> 57g plain flour
> 280ml milk
> 2 egg yolks
> Salt, pepper, nutmeg

In a heavy saucepan, melt enough lard, or butter and oil, to fry the onions. (This may be 2 or 3 tbsp – enough to ensure that bits of onion do not catch on the pan and burn.) Add the onions, and a little salt; cover the pan, and cook on a very low heat. When, after about 20 minutes, the onions have wilted and are swimming in water, uncover the pan to allow the liquid to evaporate. Cook until you have a golden, sweet mass. (Menus sometimes, a little pretentiously, call this an onion confit.)

To repeat the béchamel recipe: melt the 57g butter in a small saucepan. Add the flour, and cook for a minute, until you have a sandy (but not dark) roux. Still on the heat, add the milk in stages, stirring to incorporate each pouring before you add the next. You will have a double-thick sauce, of the sort you would use in the soufflé or cheese tart above. Season with salt, pepper and nutmeg. Allow the sauce to cool before stirring in the egg yolks (you don't want them to scramble).

Stir the onions into the sauce, then pour the mixture into the pastry case. Bake in the centre of a gas mark 3/160°C oven for about 25 minutes, until the filling is set and golden.

This tart often includes bacon, which I would treat in the same way as the bacon in the quiche Lorraine, (p95). I must admit that I have never tried the method recommended in the book: pouring boiling water over the bacon strips or lardons, draining them, and then adding them to the filling.

RICE,

PAsTA

&
COUSCOUS

RICE

Of the many aspects of cuisine that can make us feel inadequate, surely the most humiliating is an inability to prepare a palatable dish of a simple, staple foodstuff. Yet who hasn't produced soggy, clumpy, gluey rice? I have, and still do. There is one, simple rule, which – like the one about keeping your eye on the ball when playing tennis – is easier to break than you might expect. Don't overcook it.

HOW TO COOK LONG-GRAIN RICE

Method 1: bring a large pan of water to the boil. Add salt to taste, if you like.[1 – SEE WHY YOU DO IT] Throw in your rice (about 75g is a decent portion for one). Bring back the water to a simmer. Start tasting the rice at about 8 minutes. It will probably be ready in 10.[2] Drain it when it is tender but before it goes soft.

Method 2: put the rice into a measuring jug, to gauge its volume. Pour it into a pan – one in which it forms a shallow bed.[3] Cover it with twice its volume of cold water,[4] salted to taste. Bring the pan to the boil, and check the time. Simmer gently, until the water is nearly level with the surface of the rice – 'sink holes', like the bubbles you get when a thick stew is simmering, will appear among the grains. Cover the pan, put a heat disperser between it and the ring, and turn the heat to its lowest setting.

Turn off the heat 10 minutes after the water started to boil. Let the rice stand in the covered pan for a further 5 minutes.

Method 3: put the rice into a measuring jug, to gauge its volume. Pour it into a pan, as above, but now cover it in cold water and leave it to soak for 30 minutes or longer.[5] Drain. Tip the rice back into the pan, and cover it with one and a half times its volume of cold water, salted to taste.[6] Bring the pan to the boil, cover it, put a heat disperser between it and the ring, and turn the heat to its lowest setting. Turn off the heat 10 minutes after the water started to boil. Let the rice stand in the covered pan for a further 5 minutes.

A SIMPLE RICE LUNCH OR SUPPER

HOW TO MAKE IT

Fry leftover bits of roast meat in sunflower or olive oil. Add some chopped onion and/or garlic, and/or green or red peppers, if you like, and continue cooking until the onions are golden and the peppers are softened. Stir in cooked rice. Season to taste. Cooked fresh or frozen peas are a nice addition. Mushrooms, too – softened with the onions and peppers. You might also try lardons of pancetta, or bacon, instead of the cooked meat. I like to have chilli sauce on the table; others might prefer soy, or ketchup.

Method 1 is the boiled rice prevalent in Indian cuisine. We're often told that the 'absorption method' – 2 and 3, above – is the way to guarantee separate, unsticky grains. In my experience, it is not so easy to manage. Certain rices (including Tesco's basmati, I have found) become sticky more readily than others, and, unless you judge the timing water/rice ratio to absolute perfection, will settle in the pan as compacted lumps. Provided you get the timing right, boiling is foolproof.

However, it is not ideal. The more water you use to cook rice or vegetables, the more nutrients you lose from them. Boiled rice is less fragrant or tasty than rice cooked by the absorption method.

1 • Salt. Some people insist that you should never use salt when cooking rice. The edict appears to be a matter of taste rather than of science: I have never known salt produce any ill-effects on the cooked grains. Raising the boiling temperature of water, salt may speed the cooking process.

2 • Ten minutes. I am amazed at the number of recipes that tell you to cook basmati rice for 20 minutes or longer. I have yet to find a brand that is not ready in 10 minutes. Ordinary, less delicate and fragrant long-grain rice may take longer. Overcooked rice will be sticky.

Methods 2 and 3 are both versions of the absorption method. The rice boils for a while, until it has absorbed the water; then it steams, a process that completes the cooking and, in theory, separates the grains.

3 • No pre-washing or soaking. The rice you buy in packets is clean. It's harmless. In the US (though not in the UK), most of it is better for you in an unwashed state, because manufacturers coat it with vitamins that washing would disperse.

The one reason you might want to wash rice is to get rid of the polishing dust that is sometimes still present, and that turns into gum on cooking. But many brands do not require this treatment: Tilda, for example, tells me that its rice leaves the factory dust-free. A reason you might be *told* to wash rice is to rinse away the starch. If you do that, you will not have much food left: starch makes up about 80 per cent of a rice grain.

Having said that, I should add that I usually give my rice a quick rinse. Superstition, I suppose.

4 • How much water will my rice absorb? The 2:1 ratio usually works for me, and seems to give me enough water for the method I describe, which involves about 5 minutes of boiling, in an uncovered pan, 5 more minutes of boiling/steaming in a covered pan, and a further 5 minutes of steaming in a covered pan with the heat turned off. But I cannot be dogmatic about it. The size and quality of your pan, the temperature of your hob, the hardness of your water, and, most importantly, the quality of your rice are all factors that will affect the absorption rate.

Do not expect the packet instructions to be reliable. Following the instructions on a packet of basmati rice I bought in France, I cooked the rice in three times its volume of water: that proved to be far too much. In *The Cook's Encyclopaedia*, Tom Stobart writes: 'I once bought a very costly Saudi Arabian variety of rice from a Bedouin trader and found to my amazement that it absorbed no less than nine times its volume of water and would not cook properly with any less.' So, if you are tempted to buy rice from a Bedouin trader rather than from the local corner store, be warned.

However, the 3:1 ratio works if you boil the rice. Bring the water to a simmer, tip in the rice, return to a simmer, and cook for 10 minutes. By this time, the water will probably be level with the surface of the rice. Drain the water that is left. Hold the rice, if necessary, by returning it to the pan, and covering it.

5 • Soaking. Soaking elongates the rice grain, softening it and allowing water to penetrate more easily. 'This means that the grains don't break up in the pan, and therefore don't stick to each other,' Sri Owen writes in *The Rice Book*.

6 • Pre-soaked rice absorbs less water when cooking. I pre-soak rice when I make a pilaf (see p113). The volume of the other ingredients, and their ability to absorb water, invalidate the standard water/rice ratios. But if you pour in just enough water to cover everything, you can be reasonably confident that the rice, if it has already been soaking for a while, will soften when cooked.

I've assumed so far that you want separate grains of rice. Sometimes, you might want it sticky. Separate grains may be aesthetically pleasing, but they're not easy to manipulate with chopsticks. Thai fragrant or jasmine rice is stickier than rices sold as long-grain, basmati or patna. You might also try a short-grain rice sometimes sold as glutinous (sticky) rice. It is useful for rice cakes or sushi, or for those rather cloying sweet dishes that feature in Asian cuisine.

RISOTTO

HOW TO MAKE IT

Generous quantity for 2

1 litre chicken stock
1 small onion, or 2 shallots
30g butter

200g risotto rice[1] – SEE WHY YOU DO IT
25g to 30g grated Parmesan

Heat the stock to simmering point, on the ring next to the one on which you'll cook the risotto. Chop the onion finely and, in a heavy pan,[2] soften it in half the butter (see p164). You want it to be thoroughly mellow, so you may need to cook it for 15 minutes; a shallot will take less. Add the rice, and turn it with the onion and butter until the grains turn milky. Now add a ladleful of stock; stir the rice and stock, which should be simmering gently, until the rice absorbs the liquid. Then add another ladleful, and wait for that to be absorbed before adding the next.[3] Keep stirring. Some rice will stick to the bottom of the pan; don't worry about prising it loose and stirring it back into the risotto.

You'll be able to tell when the rice is nearly ready: it will plump up. Getting to this stage may take between 18 minutes and half an hour. Start tasting; you want the rice to be *al dente* (having a slight firmness to the bite), with a chalky quality. When it has got to that stage, turn off the heat (the risotto will continue to cook for a while, so the rice shouldn't be soft already), and add the rest of the butter, along with the Parmesan. Put a lid on the pan. Leave for 2 minutes, to allow the butter and Parmesan to melt; stir them in. Taste for salt; you may find that the Parmesan has contributed all the salt you want.

VARIATIONS

That is a very basic risotto; but all others, with the possible exception of seafood ones (see p109), are cooked in this way. You merely add the ingredients you want, but you need to know when to add them.

Risotto alla milanese, officially, includes beef marrow. Not many people use that nowadays, but they do use saffron – one little sachet, or half a teaspoon of strands dissolved in stock (which, officially, should be beef or veal stock), for 4 people. Add it towards the end of cooking; longer cooking diminishes its impact. You also include some wine, white or red; it goes in after you've coated the rice, and is cooked until it evaporates or the rice absorbs it (see p106), before you start ladling in the stock. I recently used a glass of white wine in a risotto for 4 people; it turned out to be too much. Anna del Conte recommends 6 tablespoons.

Other alcoholic options are vermouth, sherry and champagne. For an even richer finish, add some cream at the end of cooking instead of the butter.

Bacon or pancetta risotto is good. Fry some cut-up bits of streaky bacon, or some lardons, or cubes of pancetta, until they are turning crispy, before adding the onion (if you put the bacon in second, it won't go brown); some garlic would be nice too. Bacon goes very well with green vegetables, which you should cook apart and add to the risotto at the end of cooking, just to warm through, in order to retain their freshness of flavour. Boil some fresh or frozen peas in the stock (don't drain the stock down the sink when they're ready; pour it back into the stockpot); fry slices or matchsticks of courgettes; steam broccoli or broad beans; roast asparagus. If you've fried the vegetables, you probably don't want the extra knob of butter from the basic recipe.

Mushrooms will stand up to longer cooking, so you can add them to the onions before the rice goes in. Dried porcini mushrooms (see p162), in addition to or instead of fresh ones, contribute an intense fungal flavour: soak them in warm water for half an hour, then add them to the pot at an early stage too, along with the flavoured liquid in which they've soaked. Try also cubes of squash sautéed with the onions; or roast them and add them at the end.

WHY YOU DO IT

1 • For a risotto, use risotto rice. It is worth being a purist and buying an Italian rice, such as the widely available arborio, for your risotto. Some aficionados insist on carnaroli. Anna del Conte reports that the Veneti make their seafood and vegetable risotti with vialone nano. You generate the starchy consistency of risotto by frequent stirring. The Italian varieties will absorb a great deal of liquid, and endure a great deal of stirring, without breaking up.

2 • Thick pan. You need a thick-bottomed pan. It doesn't matter if the risotto sticks a little, and if the stuck bits are stirred back into the rest; but in a thin pan the stuck bits will burn. You might use a thin pan and a heat disperser, but that arrangement will hamper you somewhat as you stir, and will mean that adjustments of heat take a while to be effective.

3 • Ladleful by ladleful. There are two reasons for this instruction. If you use too much liquid, you risk overcooking rice that should, in the finished dish, be *al dente*, with a chalky quality; and the stirring of rice in a little liquid produces the creamy consistency that risotto-lovers prize. Nevertheless, I am in heretical agreement with Richard Ehrlich, who in *The Perfect* ...makes a case for adding a decent quantity of liquid at first. Then, after stirring it about a bit, you can leave the pot for long enough to allow you to grate some cheese, or prepare a salad, before the stock is absorbed and you need to add more. The important point is to be cautious when adding liquid after a quarter of an hour or so, because you should stop cooking when the rice is plump and at the *al dente* stage. Keep tasting it. If you run out of stock, use boiling water from the kettle.

SEAFOOD RISOTTO

Claudia Roden (in *The Food of Italy*) says that in the Veneto, where seafood risotto is a speciality, they cook the rice in plenty of stock, rather than adding liquid gradually. I haven't come across that observation elsewhere, although I can see that the less unctuous consistency that this method produces would suit fish and shellfish well.

HOW TO MAKE IT

You add the rice to a pan of simmering fish stock (see p57) and simmer it until it is *al dente*, hoping that the moment of readiness coincides with a reduction of the liquid to your required consistency. If there's too much liquid, drain some off when the rice is ready; if there's too little, add some more hot stock or some boiling water. Meanwhile, you cook some mussels (see p304): a reduction of the mussel liquor, consisting of wine and juices from the shellfish, goes into the pot with the rice. Soften some onions (see p164), and fry some prawns with them. (If the prawns are uncooked, that is: fry them in their shells, and wait for them to go pink. If the prawns are cooked, simply warm them through in the risotto.) When the rice is ready, you stir in the mussels, prawns and onions.

The best fish for risotto is a firm variety such as monkfish, which won't break up when you stir it into the rice. Add it to the pot about 10 minutes before you think the rice will be ready.

In *Gammon and Spinach*, Simon Hopkinson offers a short cut for those without the wherewithal to make a fish stock: use a bottle or two of fish soup. You could make a very nice, straightforward dish by preparing a risotto as in the basic recipe above, with hot soup substituting for the stock, and by adding towards the end of cooking some thawed frozen prawns.

The one point on which all writers are agreed is that you should never add cheese to a seafood risotto. It is an unpleasant combination. The Italians don't add cheese to seafood pasta dishes, either.

PAELLA

HOW TO MAKE IT

For 4

20 mussels

glass of white wine

4 chicken thighs or drumsticks

4 chorizos (the uncooked sausages, not the salami-style stuff)

1 onion, chopped

2 red peppers, deseeded and cut into strips

2 garlic cloves, finely chopped

400g rice (a Spanish one if you can find it, or an Italian risotto rice)

1.25 litres chicken stock

20 prawns

1 sachet of saffron, or 1/2 tsp saffron threads, dissolved in stock

Handful of parsley, chopped

Steam the mussels (see p305) with a glass of white wine. Remove the opened ones from the pan, and boil the liquor until it has reduced by half.

In a little olive oil and at a moderate heat, fry the chicken pieces, in a wide pan that you can put into the oven (perhaps a wide casserole dish, or a frying pan with a detachable handle).[1] – SEE WHY YOU DO IT After cooking for 5 to 8 minutes on each side, they should be nicely browned. Remove them to a plate, and chuck away the oil.[2]

Add a little more oil to the pan, slice the chorizos thickly, and fry them. They should release their own, paprika-flavoured oil. When they have done so, and when they are browned, take them out of the pan and put them with the chicken.

In the same pan, soften the onion, peppers and garlic; add a little more oil if you need it (but there should be enough). The peppers will produce some liquid; let this cook away. Tip in the rice, turn it with the vegetables, and cook it until it turns milky. Pour in the stock, and bring the pan to a simmering point. Cover the pan tightly with foil (or, if you're using a casserole, with its lid), and put it on to the bottom shelf of a gas mark 4/180°C oven. After 30 minutes, take it out of the oven, uncover it, and add the chicken, chorizo, mussels, prawns, saffron and parsley, stirring them in gently. Cover again, and put back in the oven for 15 more minutes.

VARIATIONS

Other meat and fish you might add include ham, snails (you can buy vacuum-packed ones) and clams (you prepare and cook them as you do mussels – see p304). Other vegetables include deseeded and chopped tomatoes (see p194), peas, broad beans and green beans. The clams, peas and beans will need cooking before you add them, after 30 minutes.

WHY YOU DO IT

1 • **You don't have to be authentic**. This dish makes me nervous. Clearly, I lack the essential machismo – in Valencia, home of the most famous recipe, making paella is the men's job. (I'm also short of confidence when faced with that other favourite masculine task, barbecuing.)

Here's what gets me worried. A real paella pan – a caldero, which is wide and shallow with handles, or the wide frying pan that you will need as a substitute – is much wider than the hotplates on cookers. The rice in the centre of the pan cooks more quickly than that at the outside, and may well stick and burn. The pan, in most recipes, is uncovered, so the liquid is absorbed into the rice and evaporates as well. After 20 minutes or so, you find that you have some cooked rice and some that has scarcely developed from the state it was in when packeted. You're not supposed to stir it, though; the dish should not resemble a risotto. So you add more water, overcooking the central portion while you try to bring the rest to an edible condition.

My compromise is to use a covered pan, and to put it in the oven. Because you're going to leave it undisturbed, you need to be confident that you've added the right quantity of liquid to cook the rice and be absorbed by it. I hope that my suggestion – 1.25 litres of liquid to 400g of rice – works for you.

2 • Throw away the chicken fat. Because, by the time you've fried the chorizo, you should have plenty of fat, and it will be nicer than the stuff the chicken exudes. If you fry the chorizo first, and then the chicken, you will probably have too much fat in the pan; and, if you discard any, you will lose some of the chorizo's paprika-flavoured oil. Don't throw the fat down the sink: it will congeal. I pour it into a saucer, then scrape it when cold into the bin.

Pilafs

Basic pilaf. This is method 3 (p102), with the addition of spices. Alternative spellings, often found in your local Indian restaurant, include pilau and pulao. Before you add water, you turn the soaked and drained rice in oil, butter or ghee (clarified butter), with flavourings that might include cumin, coriander, cardamom, turmeric, allspice or cinnamon. If they're finely ground, the spices will burn easily, so just turn them with the rice briefly.

Fancier. Richard Whittington (*Home Food*) has a recipe for turmeric pilaf with raisins and flaked almonds. Adapting his recipe for 4 people, I'd recommend 220g long-grain rice, stock or water to cover the ingredients (see point 6 on p105, above), an onion, a pinch of allspice, a cinnamon stick, 4 cardamom pods, a tablespoon of raisins, a bay leaf, salt and 20g toasted flaked almonds. Slice the onion, soften it in oil or butter, then increase the heat and add the rice, turning it until it becomes milky; add the spices, turning them quickly without burning. Add the stock or water, cinnamon, cardamom, raisins, bay leaf and salt (a quarter of a teaspoon, perhaps). Bring to a simmer, cover and cook as in the basic recipe (p102). When the rice is ready, discard the bay leaf and cardamom. Whittington recommends stirring in a knob of butter. Scatter the almonds on top before serving.

Meat pilafs. Sri Owen has an Afghan recipe called Qabili pilaf. You make a basic stew (see p229) by frying lamb pieces or chicken pieces followed by an onion, then adding salt and water, and covering and simmering for 40 minutes (the chicken will be ready in that time; the lamb may need longer). Drain the liquid, which will become the stock in which you cook the rice. Add some carrots, cut into matchsticks, to the meat and onions, as well as some raisins. Add, too, some cumin and saffron. Put the pan on the heat again, add the rice and turn it with the other ingredients; pour in enough stock just to submerge the ingredients, bring to the boil, turn down the heat, cover and simmer as above. For 4 to 6 people, Owen suggests 450g rice, 800g lamb or 1 jointed chicken (or 6 chicken portions), 1 onion, 2 or 3 carrots, 112g raisins, 2 tsp cumin and 1/4 tsp saffron.

PASTA

As pasta is now the default, easy, everyday meal of choice in so many households, offering advice on such matters as the size of the cooking vessel, the vigour with which the water should boil, and the addition to the water of salt may seem as affected as giving a detailed recipe for fish fingers – or as telling you, as I have already done (p78), how to boil an egg. Still, I shall risk giving offence.

I should guess that of all the recipes currently finding publication, pasta sauces occupy the largest category. The demand is so great that writers are under constant pressure to come up with new dishes, most of which are variations on ones below or elsewhere in this book.

Fresh pasta

It's fun, and satisfying, to make your own pasta; but it takes quite a lot of fiddly work, and requires you to clear half an acre of space to accommodate the drying pasta sheets. So I'm going to cop out and say that making your own pasta is a more committedly foodie operation than this book is designed to describe. Also, perfectly helpful instructions come in more specialist books (try *Marcella's Kitchen* by Marcella Hazan) or with pasta-making machines.

Dry pasta

What most of us use, at least 95 per cent of the time.

HOW TO COOK IT

Bring a large pot of water to the boil.[1 – SEE WHY YOU DO IT] Add salt,[2] and return to the boil. Tip in the pasta,[3] stir it, and cover the pot with a lid until the water comes to the boil again. Uncover, stir, and continue to cook at a rolling boil[4] until the pasta is *al dente* (with a slight firmness as you bite it).[5] Drain, and mix immediately with sauce, or with oil or butter.[6]

WHY YOU DO IT

1 • Large pot. Pasta, though not quite as demanding as rice (see p101), is not entirely foolproof: it sticks together easily, and goes soggy and indigestible if overcooked. Giving it plenty of water to move around in – about a litre for each 100g pasta – is the first step in keeping it separate. The water will come to the boil faster if you cover the pan.

2 • Salt. You don't have to add it after the pan boils, though the water will come to the boil faster if you do, because salt raises the boiling temperature. You can be generous, using a teaspoon of salt, say, for each litre of water. It flavours the pasta and reduces stickiness.

3 • What pasta, and how much? The best Italian brands are a lot better than anything else. De Cecco is a particularly good one. Don't use the quick-cook stuff. About 125g for each person is a decent, main course quantity.

4 • Rolling boil. A misguided pedant once wrote to the *Guardian* to tick off Richard Ehrlich, who had instructed readers to ensure that pasta water boiled rapidly. There was no difference in temperature between boiling water and fast-boiling water, the pedant wrote. True: but the temperature is not the point. The point is the water's agitation, which helps to keep the pasta separate.

If you put a lot of pasta into a little water, it will lower the water temperature to one considerably below boiling point, and might cause clumping. Even when you use a lot of water, it's a good idea to give the pasta an initial stir to encourage it to separate. Putting the lid on the pot will get the water back to the boil more quickly.

5 • *Al dente*. Packet instructions on cooking times are helpful, but they cannot be precise: kitchen conditions and water qualities vary too much. Appearance is a good guide: pasta turns milky white when it is ready. But it's best to remove a piece with a slotted spoon, stick it under the cold tap for a second, and bite into it. Stop cooking when the pasta has passed the crunchy stage but still has some firmness in the centre. Its residual heat will carry on cooking it for a while after it is drained.

You might think that you'd prefer your pasta to be soft; but if you aim to get it that way, you'll find it has a soggy starchiness that is hard to digest.

6 • Sauce it quickly. A cooling mound of drained pasta in a colander or bowl will stick together. Lubricate it by stirring in your sauce, or some butter or oil, as quickly as possible. When you drain the pasta, put the pot underneath the colander before all the water has drained through, in order to catch a little of it. You can use this cooking liquid if the pasta, once dressed with sauce, seems a little dry.

Pasta shapes

The ones I use most often are spaghetti, penne (the quills: I usually get the ridged – *rigati* – ones) and conchiglie (shells). Spaghetti goes well with meat and tomato sauces, as everyone who has been through the cookery rite-of-passage of making a spag bol knows, but not so well with creamy ones, in which it is more likely to stick together. Penne and conchiglie go with anything. I've slightly gone off farfalle (butterflies) and fusilli (spirals), parts of which overcook while others remain hard. I buy macaroni for macaroni cheese, that insult to Italian gastronomy (see p127).

Pasta sauces

Tomato sauce: see pp196–7. The quantities I've given are inauthentic, I'm afraid: Italians use just enough sauce to provide a coating and flavouring. But then the way we usually eat pasta, as a substantial main course rather than as a *primo piatto* (first course), is inauthentic too.

Other vegetable sauces: for ideas about dressing pasta with vegetables, see the Vegetables chapter (p137), and in particular asparagus (p139), aubergines (p140), broccoli (p147), cauliflower (p153), courgettes (p156), mushrooms (p162), peppers (p172) and spinach (p193). Some of these vegetables – the non-green ones, and courgettes – might be combined with a tomato sauce.

Bolognese sauce: see p256.

CARBONARA SAUCE

One of my favourite things. In spite of the advice of Rose Gray and Ruth Rogers (in the first *River Café Cook Book*) that this sauce goes best with penne or a similar shape, and in spite of what I myself say (above) about spaghetti and creamy (eggy, in this case) sauces, I still like this with spaghetti. The egg seems to scramble more satisfactorily, for some reason; with penne, I often end up with pasta in a runny, eggy and somewhat unappetizing liquid.

HOW TO MAKE IT

For 2
250g spaghetti
2 eggs

1 tbsp double cream (optional)[1] – SEE WHY YOU DO IT
Pinch of nutmeg
100g pancetta, cubed[2]
1/2 garlic clove, finely chopped[3]
4 heaped tbsp (about 25g) grated Parmesan, or a
    combination of Parmesan and Pecorino

Put a large pan of water on the heat for the spaghetti. Beat the eggs lightly, as you would for scrambled eggs (see p82). Add the cream, if you're using it. Season with a little nutmeg, and pepper if you like; you may not need more salt, because of the salty pancetta and cheese. Put the pancetta with no more than 1 dstsp olive oil in a heavy-bottomed pan, large enough to contain the spaghetti when it's ready.

When the pasta water is boiling, add salt, return to the boil, then add the spaghetti. (Some of it will probably stick out of the water. Allow it 20 seconds or so to soften, then stir it in.) Turn the heat under the pancetta to low/medium, and fry it gently; you're aiming to get it ready when the pasta is cooked, so adjust the temperature accordingly. About 3 minutes before the pasta is ready, add the garlic to the pancetta, and continue frying on a very low heat, with a heat disperser under the pan if necessary.

When the pasta is *al dente* (see p116), drain it, reserving a little of the cooking liquid. Tip the drained pasta into the pan with the pancetta. Turn off the heat under the pan, and pour in the eggs, stirring and tossing thoroughly; they should scramble in contact with the hot ingredients. Now add the cheese, and toss that in thoroughly.[4] If the mixture is too dry and sticky, moisten it with a little of the cooking liquid. Serve, with more cheese at the table if you like.

WHY YOU DO IT

1 • Careful with the cream. A modest portion will add lubrication to the sauce; but if you use too much, you'll stop the eggs scrambling. You may find that you prefer to leave it out, and to use a little cooking liquid if the carbonara looks too dry and clumpy.

2 • Pancetta beats bacon. I'm all for making do with ingredients to hand, but I do think, snobbishly purist as it sounds, that pancetta, with its sweet taste, works a lot better in this dish than does streaky bacon. I buy pancetta in a chunk from my local deli: it's tastier, and less watery, than pre-packaged lardons. You can use slices, cut up, as well. (You can also use cut-up slices of streaky bacon.) The pancetta will release its own fat, so use just enough oil to get the frying process started.

3 • Garlic. Fry it for long enough to soften its flavour; but if you put it in the pan at the same time as you start cooking the pancetta, you might overcook and burn it. If you want to be a bit more subtle, just let it add garlic notes to the oil: lightly crush a whole clove with a heavy knife, put it in the pan with the pancetta, keep stirring it around, and remove and discard it when it's brown.

4 • Timing. The pancetta pan should be hot, but not so hot that it immediately turns whatever egg hits its surfaces into a large, dry lump. The egg should scramble as it does when you make the best scrambled eggs, becoming creamily curdled. The 10 minutes or so that it takes to cook the pasta should suit the pancetta, too, the heat underneath which you can turn off as soon as you add the drained spaghetti. Stir the pasta around a little; then add the eggs. I find that they usually scramble perfectly under these conditions. As soon as creamy bits of egg are showing on the strands of spaghetti, throw in the cheese, which, being cool, will help to arrest the cooking.

SAUSAGE SAUCE

HOW TO MAKE IT

For 2

4 sausages (see Variations, below)

1 1/2 tbsp olive oil

1 garlic clove, finely chopped

400g tin tomatoes, drained; or 4 plump tomatoes, skinned and chopped (see p194)

142ml pot double cream

Cut the sausages into discs; don't be too concerned about cutting them cleanly. Warm the olive oil on a medium heat in a heavy saucepan, and add the sausages.

You needn't worry if they stick, or if bits of them break off, but you should stir them around if they are in danger of burning.[1] – SEE WHY YOU DO IT You just want bits of them to go brown, to add flavour to the sauce.

Add the garlic,[2] and stir it around for a minute or so. Add the tomatoes, drained of the liquid in which they were sitting but not of all the liquid inside them,[3] and the cream.[4] As you stir this sauce, you'll find that the encrusted bits on the bottom of the pan lift off. Simmer, very gently, until the sauce thickens. Check the seasoning – the sausages may already be quite salty.

I like this sauce with conchiglie: bits of sausage nestle invitingly inside the shells. I usually have grated Parmesan cheese at the table.

VARIATIONS

Chunky, sweet Italian sausages (bought from any Italian deli) are particularly delicious here. You could also use cumberlands, or fresh chorizos. Or any decent sausage you like. An alternative method is to skin them and crumble the meat, breaking it up further in the pan as it fries.

You can also buy **spicy Italian sausages**. Or throw in a whizzed dried chilli or two just before you add the tomatoes.

Leave out the garlic, if you like. When you've browned the sausage, add a chopped onion, and cook it until it softens. Then add the tomato, or just the cream; or leave them both out.

Brown the sausage on a more gentle heat, for long enough to cook it thoroughly; then add the garlic, if you like, and the cream, with a teaspoon of **Dijon mustard**. The sauce will be ready in another minute or so, when the cream has thickened.

WHY YOU DO IT

1 • Don't worry about shapeliness. The soft sausage is hard to cut into neat discs, and will probably break up a bit as you stir it around the pan. The irregularity of the chunks, and the stray bits of sausage you find as you bite into a pasta shell, contribute to the appeal of the sauce.

2 • When to add the garlic. I've seen similar recipes telling you to fry the garlic first. That would be fine if you could keep it away, once you put in the sausage, from the hottest parts of the pan, where it would burn. That's not easy.

3 • Draining the tomatoes. The whole can, juice and all, plus the cream, would give you an awful lot of sauce. I like to keep the juice that the tomatoes hold, though. If you don't want it, you can wring it out of the drained tomatoes with your hands; or, if you're using fresh tomatoes, cut them in half and scrape out the seeds and jelly.

4 • When to add the cream. You could add it once the sauce has reduced, and then simmer further until the sauce thickens again. But, if you're cooking it anyway, you might as well add it with the tomatoes. It will have a tenderizing effect on the meat.

ANCHOVY, GARLIC AND CHILLI SAUCE

HOW TO MAKE IT

For 2

2 slices of bread, crusts removed, for breadcrumbs

3 tbsp olive oil

1 small (50g) tin anchovies, drained

1 garlic clove, very finely chopped

1, 2 or more dried chillies, whizzed in a coffee grinder or small electric mill

Put on the water for the pasta (see p114). Whizz up the bread in a food processor or small electric mill; mix 2 tbsp of the resulting crumbs with 1 tbsp of the oil, lay out on a baking tray, and put in a gas mark 4/180°C oven. If they are brown before the pasta is ready, turn off the oven but leave them inside, or turn down the oven to its lowest setting, and put them on the bottom.

When the water is boiling, throw in the pasta. You can make the sauce in the time it takes the pasta to cook.

I prefer to use my own olive oil, distrusting the stuff in the tin, even though it is infused with anchovy. Put the 2 tbsp oil on a gentle heat, and add the anchovies and garlic, stirring the anchovies to break them up. They should melt; once they have done so, add the chilli, and turn off the heat.[1] – SEE WHY YOU DO IT

Cover the pan if the pasta is not ready yet.

Drain the pasta, and toss it with the sauce. Tip in the breadcrumbs, and toss them gently too.[2]

This sauce is good with spaghetti. I like it even better with spaghettini (the finer stuff), or with linguine (fine, flat pasta ribbons).

VARIATIONS

If I have it, I like to add **parsley** (a tablespoon) to this sauce. Basil or mint are other possibilities.

The classic southern dish is **pasta aglio e olio e peperoncino** (garlic, oil and chilli) – as above, but without the anchovies. Or just aglio e olio – without the chilli. The breadcrumbs are optional – but you're not supposed to serve cheese instead. The dish should be basic, and cheap, and the flavour of the oil, garlic and chilli should be uncorrupted. You definitely should not serve cheese if you're using anchovies, because cheese does not combine happily with fish in either pasta or risotto.

You may find the flavour of a whole tin of anchovies too powerful in a dish for two. Use half a tin; or soften the flavour, and remove some of the salt from the anchovies, by soaking them in milk for half an hour before you cook them.

WHY YOU DO IT

1 • Brief cooking. You're cooking the garlic just long enough to soften some of its rawness. It will still have a strong flavour. If you eat this dish at lunchtime, be aware that you'll probably be tasting the garlic – and exuding it – for the rest of the afternoon.

2 • Crisp breadcrumbs. Add them at the end. They will take up some of the sauce, but they shouldn't go soggy.

LASAGNE

If you can find lasagne that requires pre-cooking before you assemble the dish and bake it, and if you can be bothered to boil a pile of pasta sheets and to make the effort to stop them sticking together once drained, buy it. Most lasagne on sale advertises itself as requiring no pre-cooking, and it is not as satisfactory as the other kind. It tends to overcook in the centre of the dish, and become flabby and sticky; towards the outside, sections of some sheets may remain dry and brittle.

You're supposed to be generous with the sauce, in order to ensure that the pasta can cook, but you may find the quantities hard to judge, and end up with a very sloppy dish. But if you don't use enough sauce, the sheets won't soften. The answer is to ignore packet instructions and cook the dish at a moderate heat, covered, before browning the top.

HOW TO MAKE IT

For 4
400g lasagne sheets
Bolognese sauce (see p256)
400ml béchamel sauce (see p46), with a pinch of nutmeg
1 heaped tbsp Parmesan cheese

Spread a layer of béchamel along the bottom and sides of an oven dish, to grease it. Put in a layer of lasagne, breaking pieces to fit if necessary (if you're using the non-pre-cook variety; you'll be able to cut the pasta if you're using the other kind). Add a layer of Bolognese, then pasta, then Bolognese, then pasta, then Bolognese; that should use up the Bolognese. Put a last layer of pasta on top of it (you may not have needed all 400g), and cover the whole lot with the béchamel.[1 – SEE WHY YOU DO IT] Sprinkle the Parmesan on top. If you're using non-pre-cook lasagne, cover the dish tightly with foil, and bake in a gas mark 4/180°C oven for 40 minutes.[2] Test whether the pasta is cooked by inserting a knife. Brown the dish carefully – the Parmesan will burn readily – under the grill. If you're using lasagne that you have pre-cooked, bake the dish at gas mark 6/200°C for 25 minutes, or until the top has browned.

1 • Layering. Most recipes tell you to put some béchamel on to each layer of Bolognese, but I prefer to retain more sauce for the topping.

2 • Timing. The packet instructions on non-pre-cook lasagne may tell you that the pasta will be ready in 30 minutes, or even 20. It might be, if you drowned it in sauce. If you want the dish to have a more pleasing consistency, you need to get steam to do the work. Hence the foil covering. The process will take longer, though.

When you boil the lasagne sheets first, all the components of the dish are cooked when you assemble it. The baking is simply a matter of merging the flavours and browning the surface.

MACARONI CHEESE

HOW TO MAKE IT

For 2
200g macaroni
300ml béchamel (see p46)
100g Cheddar, or other melting cheese, grated
Pinch nutmeg
1/2 tsp Dijon mustard

Cook the pasta (see Dry pasta – how to cook it, p115). Meanwhile, make the béchamel.[1 – SEE WHY YOU DO IT] When it is ready, stir in half the cheese, the nutmeg and the mustard. Check the seasoning; the cheese may have provided enough salt. When the pasta is ready (leave it *al dente*), drain it, and mix it with the sauce: either in the pan in which the béchamel cooked or, if that is not big enough, in the pasta pot. Tip the mixture into a warmed, ovenproof dish, and scatter the rest of the cheese on top. Bake in a gas mark 6/200°C oven for 20 minutes, or until the surface of the dish is brown.[2]

VARIATIONS

If you don't mind a discoloured sauce, gently fry lardons of **pancetta or bacon** (100g, say) in a little oil or butter, then add the flour for the béchamel and cook the sauce in this pan. Or add the fried pancetta (but not its fat – there's enough cholesterol in this dish already) to the cheese sauce when it's ready.

Some people put slices of **tomato** into a macaroni cheese. I don't like the addition of acidic juices to a creamy dish.

Leave out the nutmeg and the mustard, if you like. If you don't like mustard, you might find it worth trying anyway: you'll be surprised by how it loses its heat in the sauce, and enhances the flavour of the cheese.

WHY YOU DO IT

1 • Pasta and béchamel. People usually describe macaroni cheese as comfort food, by which they must mean 'a wodge of something rich and filling'. Of course, Italians do combine pasta and béchamel – a floury paste with a sauce based on a floury paste – in lasagne and other baked dishes (see p125). But they make their cheese sauces for pasta with just cheese, butter and cream – a rich concoction, certainly, but not such a heavy one. Still, sometimes a good, old-fashioned, Anglo-Saxon macaroni cheese is just what you want.

2 • Baking. When you put the macaroni cheese in the oven, the sauce dries out and the pasta sticks together. If you want to avoid those results, put the dish under the grill to brown instead. But a sticky lump of cheesy pasta represents the spirit of the dish more accurately.

NOODLES

Packet instructions on Chinese noodles – the ones that come formed into rectangular shapes – tell you to pour boiling water over them, leave them for 3 or 4 minutes, and drain, perhaps rinsing them in cold water. Goodness knows why: it's a disastrous policy. The noodles sit in a sticky clump in the cooling water. No amount of rinsing seems to get rid of the starchiness.

Instead, cook them as you do pasta (see p114), stirring to separate them as quickly as possible, perhaps with a little oil in the pan;[1] – SEE WHY YOU DO IT most brands cook in 3 to 5 minutes. If you need to keep them for later, drain and rinse them, but don't leave the noodles in a lump in the colander. Separate them out on paper towels. That's fiddly, I know: so it's better to cook them just before you need them, adding them to broth, or a stir-fry, as soon as they're drained.

SIMPLE STIR-FRY

HOW TO MAKE IT

For 2

2 spring onions, tough green leaves removed, chopped, washed
1 garlic clove, chopped
1 tsp ginger, minced
Large handful of pak choi or other cabbage, finely chopped
250g cooked noodles (see above)
1 tbsp groundnut, sunflower or vegetable oil
1 dstsp sesame oil
Soy sauce

Having ignored the instructions on your noodle packet, you need also to disregard the many stir-fry recipes that tell you to heat oil to smoking point before adding your ingredients. Why would you want to eat degraded oil? Add the oil (groundnut, sunflower or a vegetable oil that can withstand high temperatures are best) to a hot pan or wok; then immediately throw in your first ingredients.

A wok will work better than a frying pan; and a good wok, seasoned (see p25), will work better than a bad one. Put it on a medium to high heat, wait for it to get very hot, pour in the oil, swirl it around, and immediately throw in the spring onions, garlic and ginger. Keep them moving rapidly around the pan; they probably won't need more than 30 seconds. As soon as the onions have wilted and the garlic has started to brown, throw in the cabbage. Continue to move the vegetables around the wok energetically. When the cabbage has wilted, add the noodles. When they are warmed through, take the wok off the heat, and stir in the sesame oil and a few shakes of soy sauce. You can throw some salt into the pan at any point; but remember that oriental sauces are quite salty.

VARIATIONS

As there are writers who have devoted not just one but several books to wok cookery, I cannot pretend to do justice to the subject here. I offer just a few variations on the theme above.

Other sauces: chilli, oyster, rice wine, fish (nam pla).

Other vegetables: what do you fancy? Cut them all into bite-sized pieces; add one vegetable at a time to the pan. After you've fried the garlic/spring onion/ginger base, put in first the vegetables that will take longest to cook.

Meat: tender cuts (chicken breast, pork medallion and so on), also in small pieces. Add them to the onion/garlic/ginger mixture. Or, use pieces of leftover meat; add these to the pan later, but make sure that you warm them through.

Seafood: shellfish, particularly prawns. Other seafood will probably break up in the pan. But you could cook it apart and serve it on a stir-fry base.

WHY YOU DO IT

1 • Oil in the water. I used to think that the advice to put oil in the pasta pot was redundant – surely the oil couldn't be effective in company with so much water? Then I found that Harold McGee (*McGee on Food and Cooking*), who has the best credentials to pronounce on such matters, recommended it, particularly for noodles. Lifting the noodles through the oil helps to separate them, he writes.

COUSCOUS

Most of the couscous we buy is pre-cooked. You can ignore the recipes that tell you to steam it above a stew for half an hour in a couscoussière (a tall, double-layered stewpot), or to improvise with a muslin-lined sieve. Ignore, too, the packet instructions, adherence to which may give you clumps of sticky grains, and try this.

Put the couscous – 100g is a hearty plateful for each person – in an oven dish. Boil a kettle, and carefully pour over just enough water to soak the grains – they should be damp, but not swimming. Cover the dish with foil, and put it in the oven – the temperature doesn't matter that much, but gas mark 4/180°C works fine – for 5 minutes. Remove the dish and uncover it, to reveal a solid mass of couscous grains. Pour over a little olive oil, or add a knob of butter, and stir it through; the grains should separate. (If you have had time to leave the dish in the oven for longer, say 20 minutes, the grains may separate more readily.) Salt to taste.

Why hot water? The grains may be advertised as pre-cooked, but they emerge resembling little pieces of grit if you soak them in cold water. The oven-heating helps them to dry out and separate when you add the oil or butter. (But if you can't be bothered with that, try soaking the couscous in a pan, with the lid on.) You could soak the grains in hot stock.

WHAT TO DO WITH IT

The classic way to serve couscous is with a **meat or fish stew** (see pp227 and 300), made with spices such as cumin and turmeric (see p253), and with plenty of sauce, which the grains will soak up. **Tomatoes** often play a part, as does **harissa**, a fiery paste that you can buy in tubes or tins, or that you can make yourself (see p134); add it to the stew, or moisten it with the cooking liquid and put it on the table, so that guests can add it as they like to their plates.

Bake **vegetables** such as aubergines (p140), courgettes (p156), peppers (p172) and onions (p164), and stir them when cooked into the couscous. Accompany this mixture with a **tomato sauce** (pp196–7), in which toasted spices – for each 400g can of tomatoes, a teaspoon in total of cumin and coriander seeds, warmed in a dry pan until giving off a toasted aroma, then ground in a mortar or herb mill – have been added to the garlic and oil base. Serve hot or cold.

Try couscous with **root vegetables**: baked chunks of sweet potato and onion; turnip and/or celeriac, braised in a covered pan in a couple of centimetres of stock (or water) with, for each

450g vegetables, a tablespoon of olive oil and a teaspoon in total of caraway seeds (toasted and ground first, if you like), turmeric and cayenne. Leave enough of this braising liquid to moisten the couscous.

For a **couscous salad** for 4: finely chop half a red onion, and soak it in (boiling or cold) water for an hour (see p164). Squeeze it dry in paper towels. Dry-fry a small packet of pine nuts in a small pan (careful, they burn easily). Chop a handful of parsley. Grate the zest from half a lemon. Mix all these with 250g couscous, with salt to taste, and a little more extra virgin olive oil, if you like. If the salad needs more sharpening, use a little lemon juice.

HARISSA

This recipe is adapted slightly from Tom Stobart's in *The Cook's Encyclopaedia*.

HOW TO MAKE IT

Soak 30g dried red chillies in hot water for an hour. Finely chop a clove of garlic. In a dry pan, heat 1 tsp caraway seeds, with 1/2 tsp each of cumin and coriander seeds, until lightly toasted; whizz them, with the garlic, in a mill or grinder. Scoop this mixture into a bowl. Drain the chillies, and whizz them too, until thoroughly ground up – you may need to do it in batches. Mix the chillies thoroughly with the spices and garlic, and a little salt, adding some olive oil to make a paste. You can keep the harissa in a clean, covered jar in the fridge, provided that it is protected from the air with a layer of olive oil; it should stay fresh for two to three months. Half of each of these ingredients, though, will produce plenty of harissa to be going on with.

An alternative, milder recipe includes a red pepper, baked, skinned (see p172) and whizzed with the other ingredients, with the quantity of chillies reduced according to taste. Even a baked pepper, though, has a high liquid content, and will produce quite a thin-textured sauce.

Appearances aren't everything. That useful lesson is worth bearing in mind when we buy and cook vegetables; or, if we haven't learned it already, perhaps it's a lesson that buying and cooking vegetables can teach us. The supermarkets and some chefs take advantage of our willingness to believe otherwise.

Of course, fresh vegetables and fruit should have such qualities as firmness and sharp colouring. But those qualities do not guarantee good flavour. The perfectly formed, brightly packaged veg and fruit you find in the supermarket are so often less interesting to taste than the comparatively unprepossessing specimens in the local green-grocer.

Supermarket produce may lose flavour as a result of the amount of refrigeration it undergoes. Certainly, at home, you'll find that there are several items that keep better at room temperature or slightly below than they do in the chilly surroundings of your fridge: among them are aubergines, cucumbers, peppers, potatoes and tomatoes.

Dull, olive-green broccoli looks unappetizing on the plate, and has probably spent too long in the pan. But following the advice of chefs on how to preserve bright colours may sacrifice flavours and nutrients. Chefs' recipe books usually tell you to use plenty of boiling water for green vegetables, which you then 'refresh' in ice-cold water, to fix the colour; you heat them up again before serving. In my experience, vegetables treated in this way are dull to eat; and they lose more nutrients than they would if cooked in only a little water. Even three-star chefs value taste less than they do the seduction of their customers with appearance – because they know that an enticing colour will convince people that the taste is delicious.

I'm prepared to lose a little colour by steaming vegetables. The key factor to get right is the cooking time: broccoli may need no more than 3 minutes, for example. (Make sure that the water is boiling before you put the vegetables in your steamer.) Test it with the point of a knife; it should be tender but not soft.

The hardness of your water will affect your vegetables. I have noticed that green vegetables go greyer in soft water areas than they do when I cook them at home, in the hard water that comes out of London taps. It may be that some soft waters have a higher acidity, which draws out greenness, than does the stuff that Thames Water provides. Acidity slows cooking too, as anyone who has tried softening potatoes in tomato sauce will have discovered. Another technique is to put vegetables into just enough water to submerge them, and to cook them, uncovered, until the water has evaporated. It works well with broccoli, cabbage and carrots.

Steaming and water reduction are more difficult to manage if you have a large quantity of vegetables. Piled on top of each other in a pan, they will cook at different rates according to their exposure to the steam. If you're attempting the water reduction method and you have a substantial mound of carrots, you will need a lot of water to cover them; by the time it has evaporated, the ones at the bottom will have overcooked.

Salt in boiling water, Harold McGee tells us, speeds the softening of vegetables; it also reduces the loss to the water of salts and sugars.

I abandon best nutritional practice when I cook potatoes. Put in cold water and brought slowly to a gentle simmer, they hold their shape better than they do if steamed. Parboiling potatoes for roasting expels surface starch and helps to rough up the exteriors, which crisp in the fat.

Grilled and baked vegetables are fashionable, in part because we associate them with the Mediterranean, and in part because we assume, not always correctly, that they retain their goodness better than do vegetables cooked in water. In fact, an oven-baked potato loses more vitamin C than does a boiled one. Nutritional considerations apart, baking often works better than grilling, which is not efficient at getting many vegetables tender.

Asparagus

Like most vegetables, asparagus is widely available all year round. Like others, but to an exceptional degree, it is superior as native, seasonal produce. The time to buy asparagus is May or June. Restraining yourself until then has a psychological as well as a culinary benefit: you'll feel that the expense, because indulged rarely, is justified.

HOW TO COOK IT

If there is a big contrast between the tenderness of the asparagus tip and the woodiness of the thick end of the stalk, you probably won't want to eat the thick end. So there's not much point in investing in one of those upright asparagus steamers, designed to cook the spears evenly. Cut off the bits of stalk you're not going to eat and put the asparagus above boiling water in an ordinary steaming basket; check to see if they are tender to the point of a knife after three minutes, even though they may take longer. Eat hot or cold, with vinaigrette (p37), mayonnaise (p41), hollandaise (p44) or simply melted butter.

VARIATIONS

Method 2: turn in a little olive oil, season, and bake in a gas mark 6/200°C oven. About 15 minutes.

Method 3 (from Richard Ehrlich): pour a thin film of olive oil into a saucepan. Lay the asparagus spears on top, in a single layer (you may have to cook several batches). Turn the heat on the hob to low/medium, and wait till you hear the first sizzling. Cover the pan tightly and cook, without disturbance, for 6 or 7 minutes.

Baked asparagus, which has a more intense flavour than does steamed, goes well with pasta, in salads (perhaps with a few other greens, some new potatoes and some lardons) and in risottos.

Aubergines

Nowhere in the literature of cookery is the sway of received ideas more evident than in the treatment of aubergines. Routinely, recipes advise you that aubergines should be salted, rinsed and dried before cooking. You leave them, sprinkled with salt, in a colander, perhaps with a plate on top to squeeze them a little, for half an hour or so; or, more rarely, you use the technique given in *Classic Turkish Cookery* by Ghillie Basan, who suggests you soak aubergine cubes or slices in salted water. Recipes may add the gloss that the point of the exercise is 'to remove the bitter juices'.

Once upon a time, it seems, aubergines were bitter. Some varieties still are, no doubt. But I've never found one. In any event, salting probably does not remove bitterness: it disguises it (cf *McGee on Food and Cooking*).

Supporters of salting who acknowledge the lack of bitterness in modern strains of aubergine advance a second reason for the practice: that pre-salted aubergines do not soak up so much oil when you fry them. It is true that unsalted aubergines soak up all the oil you give them as soon as they touch it. But salted ones soak up quite a lot too; and, if you have cut the vegetable into cubes and are stirring them around, the flesh soon starts sticking to the pan. Stuck bits apart, the cubes are usually still firm after the 10 minutes that recipes tend to give as the cooking time. Aubergine is not nice unless it is soft.

Delia Smith says that pre-salting concentrates the flavour, ensuring that the cooked aubergine won't be watery. I experimented. I pre-salted the cubes of half an aubergine; then I baked it with the cubes of the other half, salted only as I put them in the oven. I tasted each kind. There was no difference between them.

HOW TO COOK THEM

Cut aubergines into cubes. Put them on a baking tray or in a roasting pan, and pour olive oil or sunflower oil over them – about 1 tbsp for each medium-sized aubergine. Add salt, and pepper if you like. Toss with spoons, or with your hands.

Cook at gas mark 6/200°C for 20 minutes to half an hour, or until tender.

Recipes such as Parmigiana di melanzane (see p143) and moussaka (p258) require rounds (cut 1/2cm discs horizontally) or slices (cut the aubergine in two horizontally, then cut 1/2cm vertical slices), according to your preference. Brush them with olive or sunflower oil (if you don't have a pastry brush, pour some oil into a small saucer, dip a fork into it, and brush the aubergine slices with the backs of the tines), season, and bake as above.

VARIATIONS

You can brush slices of aubergine with oil and cook them in a frying pan or a ridged grill pan, or on a barbecue. But I like the flesh of my aubergines to have a melting texture; and I find that texture harder to achieve by frying or grilling than by baking. Frying cubes of aubergine is, in my experience, tricky; but you can cook them in a covered pan with oil and onions, which exude enough moisture to stop them sticking. Soften the onions in oil first for five minutes, tip in the aubergines with some salt, stir everything around and cover. Cook on the lowest heat, stirring regularly. The aubergines should be ready in 15 to 20 minutes.

Another way of cooking aubergines is to bake them in their skins, then to mash up the flesh with garlic, oil, lemon, salt and any other flavourings you fancy. Tahini, the sesame paste with the consistency of peanut butter, works well. I like a dash of cayenne or chilli pepper. Cumin, dry-roasted in a pan until it gives off a toasty aroma and then ground in a mill or mortar, is also nice. One clove of garlic, finely chopped or mashed (p158), for each aubergine; 1 tbsp tahini; 1 tbsp olive oil; 1 tbsp lemon juice; 1 tsp cumin seeds.

There's a neat way of baking aubergines in Gordon Ramsay's *Kitchen Heaven*. You slice the aubergine in half, slash the flesh, sprinkle one half with garlic, add oil, seasoning and, in his recipe, a sprig of rosemary, put the halves together and wrap them tightly in foil, bake at 220°C (gas mark 7) for 45 minutes, then at 110°C (gas mark 1/4) for a further 25 minutes (the aim is to get the flesh really soft). You scoop the flesh (discarding the rosemary, which will have imparted its flavour) into a pan. Then, Ramsay says, you heat it to evaporate the liquid and to achieve a thick, creamy consistency. So Ramsay clearly doesn't think the juices will be bitter: this technique will *concentrate* them.

My one variation on Ramsay's method is to add the oil at the end of cooking, in order to get its full flavour.

My favourite aubergine dish is Parmigiana di melanzane. You slice and bake the aubergines (see p141), layer them in a dish with tomato sauce (see p196 – it doesn't matter whether you end up with tomato or aubergines or a mixture on top), put chopped mozzarella on top and grated Parmesan on top of that, then bake at, say, gas mark 4/180°C to blend the flavours and brown the cheese. It's delicious hot, even more delicious at room temperature, and most delicious lukewarm. For a generous amount for 4 people: 2 aubergines; 2 portions of one of the tomato sauces on pp196–8; 2 balls of mozzarella; 2 tbsp Parmesan.

Beans: green beans

The fine beans that are available all year round, and that usually come from Kenya, taste muddy and stale. As do Kenyan mangetouts. Unfashionable runner beans, with their sandpaper skins and stringy textures, are nevertheless much more vibrant in the mouth. Bobby beans, which may not look promising, can be good. The fine beans to go for, when you can find them, are the French ones.

If you insist on the brightest green beans, throw them into a large pot of salted, boiling water, cook uncovered at a rolling boil for 3 minutes (the finest beans should be ready by then, but check – they may take a minute or two longer), drain and plunge into iced water. Reheat by tossing them over heat in butter or olive oil. However, this treatment does compromise flavour and nutritional value. Steaming is better and, provided you don't overcook the beans, shouldn't cause too much loss of colour. Tossing them immediately in oil helps to fix it, too. The acid in a vinaigrette, though, will turn them grey; don't dress the beans until shortly before serving.

Not all the ingredients of a hot meal have to be hot. Green beans are very nice at room temperature.

Dried beans

It's encouraging to hear that dried beans are good for us. But enthusiasm declines when we read warnings that we have to prepare beans properly to remove toxins. It drops further on discovering that writers do not agree on what that proper preparation involves. A publishing friend, whose famously scrupulous firm has brought out a book on bean cookery, tells me that various consultants disagreed to such an extent that the editors had to settle for advising readers to follow packet instructions.

The three areas of controversy in the treatment of cannellini, haricot, kidney beans and others are soaking, fast-boiling and salting.

Soaking. Leaving beans overnight in water speeds the cooking time. You'll notice that the beans can double in size, so cover them by a good few inches. I'll confess to the fussiness of soaking and cooking the beans in filtered or bottled water: our hard tap water seems to take longer to cook them, and to toughen their skins.

Fast-boiling. If we don't mind waiting a bit longer for the beans to cook, why bother with the pre-soaking? Because, according to the Food Standards Agency, it means that a subsequent short period of fast-boiling will get rid of any toxins. Dried beans contain substances called protease inhibitors, which can block the digestion of proteins. The FSA advises that the inhibitors become inactive after 10 minutes of fast-boiling, provided that soaking has made the beans receptive to this treatment.

Red kidney beans contain lectins, which can cause nasty stomach upsets. Soaking (the FSA recommends that you do so for 12 hours), 10 minutes of fast-boiling, and proper cooking should disable these toxins.

The reason that the FSA encourages fast-boiling is that the high heat under the pan ensures a water temperature of 100°C. If you bring a pan to simmering point, then turn it down as soon as bubbles appear, the temperature may never reach a toxin-zapping level.

There's another side to the story, of course. Some people stick up for protease inhibitors, which, they say, have anti-cancer properties and prevent the formation of blood clots.

I went to an expert on proteins and legumes. Dr Claire Domoney, of the department of metabolic biology at the John Innes Centre, tells me that she does not believe it to be necessary to attempt to inactivate the protease inhibitors. Their effect on protein digestion may be minimal, and is outweighed by their health benefits. In any event, some of them are too stable to be affected by boiling.

However, Dr Domoney does agree with the advice about destroying the lectins in red kidney beans.

Salt. 'In my experience, once [beans] meet salt, they never give in,' writes Fergus Henderson (*Nose to Tail Eating*). My reason for leaving salt out of the cooking water is different. I put two lots of beans – they were organic haricots – on the hob: one in unsalted, the other in salted, water. I tasted them after an hour. Both, despite what Henderson and other writers would have predicted, were cooked; but they had different textures. The salted beans had a mealy texture, while the unsalted ones were creamy.

So, here's a synthesis of all this advice.

HOW TO COOK DRIED BEANS

100g is a substantial portion for each person. Soak the beans – it makes a worthwhile difference to the cooking time – in filtered or bottled water overnight. (If your water is soft, tap water may be fine.) Drain and rinse the beans; cover by about 4cm with fresh water (filtered if you like). Bring to the boil, and skim the white scum – it comes from the protein albumen, as does the scum on a meat stock. Turn down the heat and simmer, covered or, if the liquid is in danger of boiling over, with the lid of the pan leaving a small gap. To give flavour to the cooking liquid, add a peeled, halved onion and a couple of unpeeled garlic cloves.

If you're cooking red kidney beans, follow the above procedure, but boil them at full blast for 10 minutes before turning down the heat to simmering point.

It's impossible to predict cooking times. I cooked the

haricot beans on which I carried out the salting experiment (see above) in an hour, but I have known various types of dried beans take 3 or more.

Towards the end of the cooking time, uncover the pan to allow the liquid to reduce so that the beans are just submerged. It can moisten the beans you're going to serve, or form a stock for a soup. If you're keeping the beans for a day or two, store them in their cooking liquid.

For an accompaniment to sausages or other grills/fries (for 4): chop a clove of garlic, and soften it in 1 tbsp oil. Add 1 dstsp tomato paste and a little cooking liquid from the beans; cook this mixture for a minute or so. (Tomato paste can have a slightly metallic taste, which cooking mellows.) Add the beans and enough of their cooking liquid to get the consistency you want: moist but not soupy, I suggest. Simmer for 10 minutes, checking the saltiness and adding pepper if you like. You can serve the beans hot or at room temperature.

For cassoulet, see p247. You could also add drained beans as a garnish to beef, lamb or pork stew (see pp231 and 245).

If all this is too much fuss, buy tins. They're not bad. The liquid may be very salty, though.

Broccoli

While it is easy to find broccoli, it is not easy to find it in a bright green, firm condition. Inferior broccoli, with its tendency to become waterlogged when cooked, has the pungent, rank odour and flavour that have alienated so many victims of institutional cooking from other members of the cabbage family.

HOW TO COOK IT

It's not my experience that you have to cook broccoli in boiling water to retain its bright colour. Cut off the florets and put them in a steamer above boiling water; start to test them with the point of a knife after 3 minutes. Remove the florets when tender, and, if you like, turn them carefully in butter or oil; season.

As do other green vegetables, broccoli goes well with anchovy. For 4 people, warm a little olive oil in a pan, throw in half a clove of finely chopped garlic, let it soften for a minute, and stir in a small tin of anchovies, drained of their oil. You could add some minced fresh or dried chillies too. Keep stirring; the anchovies will melt. Turn the steamed broccoli in this mixture. It makes a very good sauce for pasta. Add a small pot of cream if you like, bubbling it for a minute or so until it thickens.

The flavour of a whole tin of anchovies will be quite assertive. You may like to use fewer. Just one will add savouriness (umami) to the dish.

VARIATIONS

Purple sprouting broccoli, available in the spring and early summer, is finer tasting than the widely available green (calabrese) variety. Because it is more slender, you can cook it with no water, or very little. Try on it the Ehrlich method of cooking asparagus (see p139); but use a little more oil, and don't bother about getting the florets into a single layer. After a minute, take the lid off, give the florets a gentle stir, and add seasoning. Put the lid back on, and steam until tender.

Or you can stir-fry. Heat a wok or heavy frying pan; add some oil, swirl it around, and throw in the broccoli. (See Stir-fry, p130.) Keep it moving, and turn down the heat if there's a danger of smoke and scorching. When it's nearly ready, you could throw in chopped garlic, and/or chilli, and/or ginger. And/or some oyster/soy/chilli sauce.

Cabbage and other greens

Overcooked cabbage is an evocative smell, but not in a good way. It conjures up the grimness of school and hospital corridors; of soggy vegetables sitting in puddles of water beside globs of gristly stew. Members of the cabbage family, sprouts particularly, must be the subjects of more phobias than any other foods.

The phobias are worth exorcising. These vegetables have bright, crunchy, sometimes sharp, sometimes appealingly bitter flavours. They go superbly with bacon and other pork products, with anchovy, with cheese, and with oriental flavourings and spices. And they are cheap.

HOW TO COOK IT

Except when making a stir-fry, I like to steam cabbage, mellowing its flavour. Strip off the tough outer leaves, trim the stalk, cut the cabbage in half, then cut each half in two. Trim off the white heart. Slice the cabbage, and wash the slices in a bowl of cold water (it makes them crisper). Put them in a steamer above boiling water for 3 minutes. If they're ready, remove and toss with butter or oil and seasoning.

Braise the quarters of cabbage in a covered pan with 1cm or so of water or stock and some butter. They might take about 15 minutes to soften. Watch that the cooking liquid doesn't all evaporate. If there is liquid left at the end of the cooking time, uncover the pan and turn up the heat to evaporate it, and stir to coat the cabbage in the butter.

Heat a wok or heavy frying pan, pour in a couple of tablespoons of sunflower, groundnut or vegetable oil, throw in some chopped garlic, then add the steamed cabbage slices, stirring quickly to coat. Add some oyster, soy or chilli sauce. Or add chopped chillies at the beginning, with the garlic, and/or finely chopped ginger.

Red cabbage

Red cabbage is one member of the cabbage family that can withstand longer cooking. It is often braised with vinegar and apples, the acidity of which not only complements it in flavour, but also prevents it from losing its colouring.

HOW TO COOK IT

Slice an onion and soften it in butter or sunflower oil, or a mixture, in a heavy casserole. Quarter the cabbage, cut out the white core, and slice it; wash the slices. Add it to the onions; add also 3 tbsp vinegar, 1 tsp brown or white sugar, 2 cloves, salt, and some nutmeg if you like. If you think that the mixture is too dry and that the cabbage is likely to catch, add a little orange juice, red wine or water. Stir well and warm up the mixture on the hob; then put the casserole into a gas mark 4/180°C oven for half an hour.

Or cook it following the main instructions above, but with some acidic content to fix the colour. Or eat it raw, which is how I like it best – in a salad with rice, raisins softened in hot water and drained, soy and balsamic vinegar.

Sprouts

HOW TO COOK THEM

Wash them, peel off the loose outer layers, and trim the bases; the popular practice of scoring crosses in the stalks is not necessary. If you have a lot of sprouts, you might find steaming inefficient. So boiling them is the answer; and the most efficient way to do that, even though it loses more nutrients, is with a lot of water. Bring it to the boil, add salt, then throw in the sprouts. Check with the point of a knife after 5 minutes; they should be tender but not soft. You could then toss them in a frying pan with oil and garlic, or simply with butter.

Or: slice them and give them, from raw, the stir-fry treatment (see broccoli, p147, and cabbage, p149). It's also a good way to treat leftover sprouts.

Other greens

They are all delicious if sliced and stir-fried (see the cabbage and broccoli variations, above). **Pak choi** has a lot of stalk; separate it from the leaf, chop it into 10p-sized pieces, and put it into the wok or frying pan a minute or so before you add the green bits.

The leaves and stalks of **Swiss chard** have to be cooked separately, because the stalks can take a while to soften. Cut the stalks into fork-sized pieces (in other words, ones of the right size to put in your mouth), and put them into a pan with 1cm boiling, lightly salted water and a knob of butter. Cover and simmer. Check after 10 minutes; they may need longer. When nearly ready, uncover the pan and turn up the heat to evaporate the liquid and coat the stalks in butter. Meanwhile, slice and chop the leaves; steam them. They should take about 5 minutes. Grease a gratin dish with a little clarified butter or oil (see Gratin dauphinois, p188). When stalks and leaves are ready, mix them, add more salt if necessary, and pepper and nutmeg if you like; put them in the dish. Pour over enough double cream, or more if you like, to cover. Grate Parmesan on top. (Or cover with Parmesan, or some other grated cheese, only.) Put in a gas mark 6/200°C oven until bubbling, with a cheesy crust: about 15 to 20 minutes.

Brussel tops are my favourite winter greens, although they can taste rather dry. Wash, slice and steam them (about 5 minutes); toss with plenty of butter, salt and pepper.

Carrots

Sliced carrots soften faster than do batons, for some reason. Boiled and then drained, they lose not only their crunchiness but also their sweetness and colour. They are dreary fare.

I have never looked back since discovering a recipe for carrots in Vichy water, even though I rarely use Vichy water when I follow it. Peel the carrots, then cut them horizontally into three or four pieces; then cut these pieces vertically. You want fork-sized pieces – ones that are single mouthfuls; cut them vertically again if necessary. Put them in a pan large enough to hold them in no more than a double layer. Pour in water to come level with the top of them, or just under – they will emit their own water. Add a knob of butter, but no salt, which might make them too soft. Turn the heat under the pan to maximum, and cook, uncovered, until the water evaporates and the carrots are coated in the butter. (Be careful that the butter doesn't start to burn on the dry pan.) Turn down the heat, and gently stir the carrots in the butter. Add salt; scatter over parsley if you like.

Or (and this is now my preferred method): cut the carrots as above, and put into a pan with 1cm water and the butter. Bring to the boil and simmer, covered, for 10 minutes. Uncover the pan, turn up the heat, and cook until the water evaporates and the carrots are coated in butter.

Cauliflower

Choose cauliflowers with white, tightly packed, firm florets. Don't cook them whole: some florets will turn mushy before others get tender. Separate the florets and steam them, checking for tenderness after 5 minutes. Like its relatives in the cabbage family, cauliflower will give you a powerful reminder of the horrors of institutional food if it's not fresh, or if it's cooked for too long.

I like cauliflower cheese with roast chicken. Make a cheese sauce (see pp46–7 – about 300ml would be about right for a medium-sized cauli) with plenty of black pepper, and also some nutmeg, to offset the bland qualities of the vegetable; add too 1/2 tsp Dijon, or 1/4 tsp English, mustard. Steam the florets, put into a warm dish, pour over the sauce, scatter grated cheese on top, and brown under the grill. I sometimes prepare cauliflower cheese in advance, cover it with foil, and warm it in the oven (gas mark 4/180°C for 20 minutes). In spite of the thickening of the sauce, the skin that forms, and the evaporation that takes place on reheating, it produces a perfectly decent result.

Cauliflower, or cauliflower mixed with broccoli, makes a good sauce for pasta. It needs a bit of dried chilli, I think; and, if you're using cauliflower alone, some colour too. Follow the recipe for broccoli with anchovy (see pp147–8); but, after frying the garlic and melting the tinned anchovies, add 1 dstsp tomato paste (for a sauce for 4), and cook it for a minute before tipping in the steamed cauliflower. You could also add a little saffron, with the liquid in which it soaked (or a sachet of the powdered stuff). Other possible additions: sultanas, softened in water for 15 minutes; pine nuts, toasted gently in a dry saucepan. Crushed tinned sardine could substitute for the anchovy.

Chicory

The point of chicory, from a culinary angle, is its bitterness. Careful, slow cooking offsets that quality with sweetness; but contact with boiling liquid wrecks the bitter-sweet balance.

Try this gratin. Trim the chicory of wilted outer leaves, cut in half vertically, sprinkle with salt, and put, cut side down, in a greased gratin dish large enough to hold the pieces in a single layer. Cover the dish quite tightly with foil – you may need to wrap it underneath; put in a gas mark 4/180°C oven. After 20 minutes, have a look: if the chicory is in danger of sticking and burning, put a little water or wine into the dish. Turn the chicory over, and cook for another 20 minutes, or until tender (it might take up to an hour). Take out of the oven, and pour over a small pot (142ml – for 2) of double cream. Top with grated Parmesan, and put under the grill until bubbling and golden. You could add 1 tsp mustard and/or a pinch of nutmeg to the cream.

Another way to do it: fry some lardons (cubes of pancetta or bacon) in a little oil, until their fat runs. Remove to the gratin dish, leaving behind the fat. Fry the halved chicory in the fat very gently, turning occasionally, for 10 to 15 minutes, until quite tender. Put the chicory in the gratin dish. Make a cheese sauce (see pp46–7), and pour it over. Put in a gas mark 6/200°C oven for about 15 minutes, or until bubbling. For a substantial dish for two: 100g lardons, 4 chicories and 300ml cheese sauce.

Chillies

They're so often disappointing. Recipes usually advise you to deseed them, on the fallacious grounds that the seeds are hot; in fact, it's the white membrane that is the hottest part of a chilli. Often the flesh that's left behind once you've discarded the seeds and membrane has no more kick than that of a bell pepper. If mildness is what you're after, look for the larger varieties; small chillies, and the squat ones that might be habaneros or Scotch bonnets, should be more powerful. If you dare, it's worth tasting them before use.

I buy tiny dried red chillies, and use them a lot, particularly in pasta sauces. They're impossible to 'crumble', as recipes instruct; I whizz them in a small electric food mill.

The Encona brand of chilli sauce is good. Don't use it in a marinade, as a cheat's jerk sauce, though: it turns bitter when cooked.

Courgettes

Boiled or steamed, courgettes are mushy and dull; but, if you sauté them, they can disgorge so much water that they boil anyway. So, according to some recipes, they require pre-salting, to draw out this water. An easier solution is to sauté them on a high heat, so that the water vaporizes quickly. Buy small courgettes: they usually have more flavour, and they are less likely to flood your frying pan.

If you want a firm, crunchy texture, cook courgette batons, not rounds. Cut a small courgette about three times horizontally; cut these pieces once or twice vertically; at right angles to these cuts, make another vertical cut.

The anchovy/garlic/chilli/cream sauce for pasta (see broccoli and cauliflower, above) is good with courgettes, too. Or, keep it simple (for 2):

Soften a chopped clove of garlic in 1 tbsp olive oil, add batons or rounds of 4 small courgettes, and sauté over a medium to high heat, stirring constantly, for about 5 minutes, or until tender. Serve, accompanied by grated Parmesan, with spaghetti or spaghettini. Add a chopped, dried chilli or two to the cooked courgettes if you like. Instead of the Parmesan, try breadcrumbs: whizz up a couple of slices of bread, crusts removed, in a food processor or small electric mill, toss with 1 tbsp olive oil, put on a baking tray in a gas mark 4/180°C oven for 10 minutes, until brown. Toss gently with the pasta and courgettes.

Cut courgettes into pieces of whatever size you like,

season them and coat with olive oil, put in a roasting pan and cook at gas mark 6/200°C. Stir them around a bit after 15 minutes. They should be tender in about 25 minutes. Cook them with cubed, oiled and salted aubergine, and stir the vegetables into couscous (p132). Or include them in a ratatouille (p173). Or toss them with pasta.

Cucumber

Pre-salting, as I've said elsewhere, is usually a bothersome and unnecessary process. But it does do good things to a cucumber. Peel the cucumber, cut it however you like, put the pieces in a colander, and lightly salt the layers. After half an hour, pat the pieces dry with paper towels. Put in a bowl, and stir in a little vinegar (1 dstsp for half a cucumber). You should find that this salad has a lovely balance of sweetness, saltiness and acidity.

Pre-salting is also worthwhile if you're going to make cucumber with yoghurt (tzatziki, or cacik), because it prevents the salad from becoming watery.

For 4

1/2 cucumber
1 pot (200g) Greek yoghurt
1 garlic clove, crushed with a little salt or, for a
     milder taste, finely chopped (see p158)
Pepper (you're not likely to need more salt)
4 fresh mint leaves, chopped

Slice or chop the cucumber however you like, salt it and pat dry as above, and mix with the other ingredients.

Most cucumbers have tough skins, which I prefer to peel.

Fennel

Fennel responds well to the kind of braising you would give to meat – gentle frying followed by cooking in a little liquid. The colouring in oil or butter (or both) and the long cooking sweeten the aniseed notes of the bulbs.

HOW TO COOK IT

Trim off the fronds (which you can use to garnish the dish), as well as any tough layers, and cut the bulbs into halves, quarters or slices – whatever you prefer. Colour them in oil or butter (or both), preferably in a large sauté pan with a lid; or use a heavy casserole or a saucepan. About 10 minutes on each side. Pour in water or stock to about 1cm deep, bring to a simmer, and cover the pan. Cook gently until the fennel is tender – about 20 minutes to half an hour. Uncover the pan and turn up the heat to evaporate the liquid that remains. Add salt and pepper.

Garlic

Round my way, every other shop has a stall of vegetables outside. Much of the produce, inhaling the traffic fumes for day after day until finally the proprietor gets rid of it, looks tired; but there's usually something worth buying. If you want garlic, though, you tend to find that all of it, in these shops and elsewhere, is in a sorry state: falling apart, wrinkled, soft and sprouting. Here is one area where the supermarkets have the edge. Tesco's garlic is consistently good.

You can get away, just, with using inferior garlic when you

fry it thoroughly and cook it for a long time, in sauces and stews. But it will be bitter in dishes that require long cooking of whole cloves (chicken with 40 cloves of garlic, p243, for example), or in preparations that depend on the pungent zing of raw garlic: aioli (p42), or garlic vinaigrette (pp37–8), or hummus (p168).

Crushed garlic is more pungent than chopped. The violence of a garlic crusher turns garlic acrid and sulphurous – its relative in the allium family, the onion, acquires similar qualities if mashed in a food processor. Halve the garlic down its length, and remove from the interior any green sprouting material, which is bitter. Chop the garlic, sprinkle it with salt, and crush it with the back of a heavy knife; or put it in a mortar and use a pestle. It will become a creamy pulp. Just a speck or two of it, dissolved in the vinegar, will flavour a vinaigrette.

Frying softens and sweetens garlic, as it does onions. That's why recipes often tell you to fry garlic first, before adding other ingredients to the pan; added later, it might not get access to the oil. The drawback is that it burns easily; once you've added the other ingredients, you have to make sure that it moves from the area of highest heat.

Baked, garlic becomes creamy and mild. It's less likely to overcook – you open the skins to find that the pulp has disappeared – if you put whole heads in the oven, anointed with a little olive oil and surrounded with foil. About an hour at gas mark 4/180°C should be right; but I'm afraid that you might find some garlic to be overcooked in that time, and some still to be hard. Or you can toss the garlic cloves in olive oil, scatter in a roasting pan, cover the pan with foil, and then bake – perhaps for half an hour to 45 minutes.

The cloves steam inside the foil. Meat can overcook when treated in this way; but garlic remains more tender than it would if exposed to the radiant heat of the oven.

The cooking time for garlic is similarly hard to predict if it is boiled or steamed. When I put a clove of garlic in the water with potatoes for mashing (pp177–8), it's usually soft when the potatoes are, but not always.

Leeks

Small, slender leeks: steam them whole. Trim off the tough green bits; slice down through the middle of each leek from the top to about a third of the way down, and make another cut at right angles to the first, so that the layers fan out; wash the leeks in cold water. The reason for the cuts is that bits of grit can penetrate down through the layers. Put the leeks into a steaming basket above boiling water, and cover; cook for about 5 minutes, or until tender in their thickest parts to the point of a knife. Drain, and squeeze out excess water with the back of a wooden spoon. Then turn the leeks in butter, and season. You can also serve them lukewarm, with vinaigrette (see p37); a lovely dish is leeks in vinaigrette with a poached egg (p79) and fried bacon or lardons. Season with lots of pepper.

I'm not keen on cooking larger leeks whole. They are tough to cut through, and they have a slimy quality. Slice them, wash them, and stew them, lightly salted, in a little butter. They'll be tender in about 10 minutes. You could steam the slices, but they'll be rather watery.

Leeks sit happily in a béchamel sauce (p46), particularly

one with cheese; pour the sauce over cooked, small leeks, or stewed, sliced ones, sprinkle cheese on top, and brown under the grill. Leeks also have an affinity with fish (see fish pie, p288).

Lettuce

The pre-bagged, mixed salads you get in supermarkets may be convenient, but they're also tasteless, I find. Further grounds for shunning them may be found in *Not on the Label* by Felicity Lawrence, who describes how these salads are processed. Buy lettuces and other salad greens (avoiding the tasteless icebergs) from the greengrocer; wash them thoroughly (you'll wash the supermarkets' 'ready to eat' stuff too, if you've read *Not on the Label*), and dry them in a salad whizzer (see p29). A vinaigrette (p37) won't adhere to wet leaves. Dress green leaves just before you're due to eat them: the oil and vinegar soon cause them to go limp and discoloured.

For a more substantial green salad, try adding chopped-up bits of blue cheese, and/or bits of walnut, and/or fried bits of bacon or pancetta (with their fat – reduce the oil in the dressing accordingly), and/or croutons. Fried croutons take up a lot of oil. Instead, I bake them (for 4): 3 slices of crustless bread cut into squares, tossed with 1 tbsp olive oil, laid out on a baking tray and put into a gas mark 6/200°C oven for 5 to 10 minutes, or until golden.

Mushrooms

HOW TO COOK THEM

For 2, as an accompaniment or snack

200g mushrooms, rinsed,[1] – SEE WHY YOU DO IT woody
    parts of their stems trimmed, thickly sliced
1 1/2 tbsp olive oil, or about 40g butter, or a combination
Salt
1 garlic clove, finely chopped
1 dstsp parsley, chopped
Squeeze of lemon juice

Warm a frying pan, or saucepan with a large base; add the oil or butter. When the oil is warm or the butter is melted, throw in the mushrooms, lightly salt them, and cook, stirring regularly, over a medium heat. They will suck up the oil; keep stirring. After a minute or so they will release their liquid; turn up the heat to let it evaporate.[2] Add the garlic and parsley, and cook for a minute longer. Turn off the heat; add a little lemon juice, and pepper if you like.

VARIATIONS

This recipe will apply to most types of mushrooms. But chanterelles contain a tremendous quantity of water, and, like tender meat, toughen up if allowed to stew. Rinse them, salt them, and put them in a saucepan, with the lid on, over a high flame. Shake the pan to prevent sticking and burning. After a minute or two, they should be swimming in liquid. Drain them in a sieve, and sauté them quickly as above.

Alternatively, you could sieve the liquid into another saucepan, boil it over a high heat until it has reduced to a syrup, and add it to the cooked mushrooms.

The garlic and parsley – a Provençal theme, sometimes including breadcrumbs – are not compulsory. They are added at the end of cooking to preserve their freshness of flavour.

Tarragon and basil are also particularly good complements. Or cook the mushrooms with spices (see p253).

Another possibility: add a small pot (142ml) of cream to the cooked mushrooms, and allow it to bubble until it has thickened slightly. That, on toast, makes a nice lunch; as does the same recipe but with sour cream and paprika. You could soften a chopped shallot or two before sautéing the mushrooms.

Baking enriches mushrooms. Turn them, whole, in olive oil, sprinkle with salt, and bake in a gas mark 6/200°C oven for 20 minutes to half an hour, depending on their size.

Fungal qualities are at their most intense in dried mushrooms. You soak them for about half an hour in tepid water, drain, then stew them in butter for about 10 minutes. Or simply add them to a stew or risotto, along with their flavoursome soaking liquid.

All these mushroom recipes might be used as pasta sauces.

WHY YOU DO IT

1 • **Washing them**. Mushrooms are mostly water, so warnings that washing them will ruin their texture are wide of the mark. I put them in a colander and give them a quick rinse under a running tap, scraping off bits of earth with a knife.

2 • Absorbing oil, disgorging water. Mushrooms have the spongy qualities of aubergines. They are similarly greedy: if you add more oil, they'll absorb that too. Be patient. The heat will force them to give up the water they contain; and, once that has evaporated, a coating of oil will reappear on their surfaces.

Onions

Why do cookery writers claim that you can soften onions in 5 minutes, and brown them in 15? They're like the people who move out to the country and insist that they can get to their London offices in less time than it took them from Shepherd's Bush. It's the need to make something sound better than it is, I suppose.

You cook onions in oil or butter to make them milder and to bring out their sweetness. If you're using them in a stew, when they will cook for a good while longer, you may need to soften them for only 10 minutes or so. But if you're making a risotto, you'll find that any raw qualities in the onions will be apparent in the finished dish unless you soften them thoroughly, until golden; it will probably take about 20 minutes. Covering the pan hastens the process, but you need to lift the lid regularly to make sure that the onions are not catching and burning. Even small traces of burned onion will cause a dish to taste bitter.

Sautéing onions in butter can cause problems, because the onions absorb the fat, and catch on the base of the pan; both the onions and the butter can burn. You can thin the butter with a little oil; or (Anna del Conte's tip), keep the mixture moist with a little water.

Browning onions, to add further Maillard flavours to a

stew (p234) or to create a tasty accompaniment for grills, can take 40 minutes. If you try to speed up the process, onion pieces that your stirring fails to disturb for a while get burned. Cook slices of onion on a moderate heat, stirring regularly; make sure there's always a layer of oil in the pan, because onion in a dry pan will catch and burn. Encouraging caramelization with the addition of a little sugar might help, but it has never made much difference when I've tried it. Browning onion slices in the oven is easier, and works well (gas mark 6/200°C); spread out the slices (or they won't colour), and make sure that there's a layer of oil between them and the roasting pan.

I slice a section of onion from the stem and root ends, peel them and discard any tough layers, cut them in half from tops to roots, and slice those halves. If I'm baking them, perhaps with other vegetables such as aubergines and courgettes, I don't trim the root, and I make sure that the chunks are all connected to it – it holds them together.

Onions for a Venetian-style liver and onions should be cooked very slowly, in a covered pan. Restaurants sometimes pretentiously describe this preparation as an onion confit. (A confit implies preservation, and is properly used of jams, or of duck that is cooked, salted and stored in its own fat.) Slice onions thinly, and put them, lightly salted, in a heavy pan with a layer of oil or butter. Cover the pan, and cook over the lowest possible heat, perhaps above a heat disperser. Have a look after 15 minutes or so, and stir if the onions are starting to collapse. Stir again every so often. The onions may take an hour or longer to become soft and sweet. If there's a lot of liquid in the pan at the end of cooking, turn up the heat to evaporate it, with the lid of the pan removed. To offset the sweetness, you could add a little vinegar at the end of cooking; add parsley and pepper too, if you like. This recipe works particularly well with red onions. (Cook the liver separately. Slice it thinly, removing the stringy bits, and sauté it quickly in butter or oil over a medium heat – just a few minutes.)

Red onions, sweeter than the normal kind, can nevertheless be harsh if served raw in a salad. Soak them first, for an hour or so, in cold water, to wash away some of their sulphur; at the end of that time, squeeze them gently, drain them, and dry them with paper towels. Or give them a more radical treatment by throwing them into boiling water for 30 seconds, then draining. I usually disarm spring onions by one of these methods too.

Peas

Asparagus takes less time to cook than most cookery writers allow; fresh peas take more, I have found. Maybe I've been buying inferior peas. Far from being ready in 10 minutes, they often retain the texture of bullets at 15. So I'm a little wary of recipes that tell you to braise peas gently in a little wine or water.

Still, braising is the best method, if your peas are small and fresh enough. Heat 1cm water, olive oil or butter and salt in a pan, with some fresh mint if you like; when the liquid is simmering, add the peas and cover. If there's liquid left when the peas are nearly ready, boil it hard, with the lid off, to evaporate, coating the peas in the oil.

Once picked, peas convert their sugars into starch. Frozen peas, having been refrigerated soon after removal from their pods, are exceptionally sweet. Their sweetness is a bit one-dimensional; there is more depth of flavour in a good fresh pea, and the slight starchiness has its own appeal.

Nevertheless, frozen peas are the best frozen vegetables. Try them puréed or mushy: add them to boiling water, bring back to the boil and simmer for a couple of minutes, drain, and mash roughly with a potato masher. Over a very low heat, above a heat disperser, stir in butter and salt. (Sorry to be vague about the quantities, but I do feel that you should use your own judgement here.)

Chickpeas

See Dried beans (p144). You'll certainly find it a lot easier to cook chickpeas if you soak them overnight; even after that, some can take 4 hours to soften. Drain them, bring to the boil in fresh water, and simmer; a peeled and halved onion and a couple of cloves of unpeeled garlic in the liquid will do no harm, particularly if you're going to use this liquid later, perhaps to thin some hummus (see p168). Cover the pan or, if the boil is too fast and the liquid is in danger of boiling over, leave the lid slightly ajar. The chickpeas are unlikely to be ready in less than 2 hours.

Bicarbonate of soda in the soaking water and/or in the cooking water will reduce the time it takes to soften the chickpeas – quite dramatically, if you use enough of it. But there is a price to be paid, in loss of nutrients, texture and flavour.

If I'm in a hurry, I open a can. Canned chickpeas may not have the earthy nuttiness of the dried variety, but are worth eating nonetheless, and make a decent hummus.

HUMMUS

For 4, as a starter
400g can chickpeas, or 150g dried ones, cooked as above
1 1/2 tbsp tahini paste
1 garlic clove, chopped and crushed with some salt
Juice of 1 lemon
2 tbsp extra virgin olive oil
Pinch of cayenne pepper or chilli powder
Salt

I have come to prefer mashing the chickpeas with a potato masher to blitzing them in a food processor; the purée may be rough, but it has more chickpea flavour. Drain the chickpeas; if using canned ones, rinse them of their briny liquid. Put in a saucepan, and mash. Stir in the other ingredients. Have a taste: you might want more lemon, or salt. If the mixture is too gungey, add some of the liquid in which the chickpeas have cooked; if I'm using canned chickpeas, I pour in a little orange juice as a thinner.

Or: put the ingredients, minus the olive oil, in a food processor, and blitz. Stir in the oil at the end. Blitzed, it loses its fruitiness – see Mayonnaise, p41.

You could flatten the hummus on a plate, drizzle over a little more olive oil, and dust with paprika.

Vary these ingredients according to taste. Some people prefer more garlic; some, more lemon. The tahini, a sesame seed paste, is optional. There are brands that are rather cloying. A tablespoon or two of yoghurt will lighten the texture.

Chickpeas are a nice accompaniment to lamb, as is spinach; and chickpeas and spinach go well together. Cook the chickpeas, and drain them; cook the spinach (see p193), drain, and chop it. Fry garlic and cumin seeds in oil. Add the chickpeas and spinach, and stir them around to flavour them. A few saffron threads with their soaking liquid would add attractive colour and fragrance; or you could add a little turmeric. For 4: 200g chickpeas, 750g spinach, 1 clove garlic, 1 tsp cumin.

A more satisfying way to cook this dish would be to add the spinach and the spiced oil to the simmering chickpeas; but, as the chickpeas are sitting in liquid and as the spinach throws out a good deal of water of its own, you'll end up with a soup rather than with a side dish.

Or: drain a can of chickpeas, add them with a little water to the fried garlic and cumin, cover the pan, and heat gently to warm through until the liquid evaporates or is absorbed (uncover the pan and turn up the heat to boil away the liquid if there's too much once the chickpeas are warmed through); add the cooked, drained and chopped spinach. Because of the salt in the chickpea brine, you may not need more.

Lentils

Lentils are humble things that are a challenge to cook to perfection. The best ones are Puy or other green lentils, followed by brown. The split, disc-like, green-brown ones have a more muddy taste, but are fine for soup. Red and yellow ones turn quickly to mush, and are good for dal.

You can put green lentils into a large pan of boiling water, cooking them for 20 to 30 minutes, after which they should be tender but still holding their shape. Then drain. But the liquid you are pouring away, or perhaps saving to use with soup, contains nutrients and flavour. The alternative is to try to use an amount of liquid that will leave the lentils, when cooked, moist but not drowned.

HOW TO COOK THEM

Because lentils cook quickly, you don't need to soak them. They do need washing, though, and picking through: some packets will contain discoloured and damaged ones, as well as extraneous material including, sometimes, small stones. For 4 people, chop an onion and finely chop a clove of garlic; soften the garlic in 2 tbsp olive oil for a minute; add the onion and soften for 10 minutes. Add the lentils, cover by about 1cm with cold water, add salt (for remarks about whether salt will toughen the lentils, see Dried beans, p144), and bring to a simmer. Turn down the heat, and cover the pan. Check on progress after 10 minutes or so: lentils absorb a lot of water, and you may need to add more. But don't add more than the topmost layer will need to continue cooking. As the lentils near softness, uncover the pan. What you hope is that most of the liquid will have evaporated at the moment when the lentils have softened but have not broken down.

Alternatively, cover the washed lentils with water and cook as above, but soften the garlic and the onion in a separate pan, perhaps for 20 minutes, until they are golden and sweet. Stir the garlic and onion into the lentils just before serving. It depends what you want: a garlicky, oniony flavour that is integral to the dish; or one that is part of it but distinct.

I have to admit that what you might call the 'absorption method', above, is hard to get right. I have taken to simmering lentils in salted water so that I can drain them when they're tender but not mushy. For a salad, I add them to a vinaigrette, perhaps with chopped spring onions.

DAL

HOW TO MAKE IT

For 2

Use red or yellow lentils (200g). Wash them, and remove any grit and extraneous material.

For ideas about spice mixes, see pp253 and 272. Or cheat and get out an Indian or Thai curry paste; I use about one and a half times the quantity suggested on the bottle. Soften a clove of chopped garlic in butter, or butter mixed with sunflower oil (or, if you have it, ghee). Add a chopped onion, and cook for 10 minutes or so, until softened. Add spices (toast whole seeds such as cumin and coriander in a dry pan, and grind them in a mill or grinder; use 1 tsp of the spices you want to be most prominent, and less of any others), or curry paste, and cook for a further minute. Add lentils, and, for a rich dish, a small carton of coconut cream (or 1 tbsp creamed coconut, which comes in a block), plus water to cover. Careful with the salt: the paste may be salty.

Cook gently, covered; be careful that the coconut cream

(you could also use a tin of coconut milk) does not stick to the bottom of the pan and burn. The lentils usually soften in 20 minutes or less. Uncover the pan, turning up the heat if necessary, to reduce the dal to a thick, gloopy consistency. If you like, add chopped fresh chillies and coriander leaves.

Most recipes tell you to cook the lentils and the onion/spice mixture apart. I simmer the lentils, covered, with turmeric, chilli pepper, and salt, with frequent checks on the level of the liquid. Meanwhile, I make the onion/spice mixture as above. When the lentils are soft, and thick and gloopy, I tip them into the mixture, stirring everything around for a minute or two.

Peppers

Raw slices of pepper in salads are indigestible, I find – the ripe red and yellow ones just as much as the green. The flesh is watery; but it's the skins that are disagreeable. They remain so after cooking in a stew as well. I like to get rid of them.

There are three ways to burn and loosen the skin: on a skewer turned above a gas flame; under a grill; and in the oven (see below). You could follow the usual recommendation of putting the charred peppers in a bowl and stretching some cling film over the top; but I've never found that this steaming makes much difference. Scrape off the skin, and put the peppers into a bowl. Cut them up, scraping out the pips (a somewhat fiddly and messy business) and keeping the liquid, which you may want to strain into your stew or even salad dressing.

The skin of grilled peppers burns before the flesh has done much cooking. For a softer, sweeter pepper, try baking.

Put the peppers on to a lightly oiled (to prevent sticking) baking tray in a gas mark 6/200°C oven for about 40 minutes, turning once; they're ready when the skin is loose and charred in places. Cut the peppers as above.

I use these baked peppers in ratatouilles and other stews. Of course, they don't need any more cooking; so if you're putting them into a stew, add them at the end, just to warm through. If you're using unskinned pepper slices in a stew, you don't need to treat them as you do onions, softening them in oil first (though you will end up with slivers of shed skin in the concoction). You soften onions, and garlic, to remove some of their harshness; but peppers are milder at heart, and cook quite happily in liquid.

RATATOUILLE

HOW TO MAKE IT

For 4

1 large aubergine

3 red peppers

2 onions

3 courgettes

400g tin tomatoes, or 6 medium-sized fresh ones,
    skinned and chopped

2 garlic cloves, finely chopped

Olive oil

This is an inauthentic recipe, but it honours in part the principle of ratatouille: that the ingredients be cooked apart, then combined.[1] – SEE WHY YOU DO IT

Cut the aubergines into fork-size chunks, and put them in a roasting pan. Toss them (with your hands, if you like) with enough olive oil to give them a good coating, and with salt. Put the peppers, whole, in the pan alongside them. Bake at gas mark 6/200°C, turning the aubergines after 20 minutes. Cook for about 40 minutes in total, or until the aubergines are tender and the peppers have blackened skin (leave the peppers in the oven for longer, if they need it). Skin and deseed the peppers when they're cool enough to handle, retaining their juice.

Peel the onions, cut them in half, and slice them thinly. Soften them in olive oil over a gentle heat for 20 minutes, or until golden and sweet. Cut up the courgettes as you like, turn up the heat a little, and throw them into the pan, stirring them until soft (see p156). (In the first edition of this book, I advised roasting the courgettes with the aubergines. I now think that they have a more vivid flavour when fried.)

Make a simple tomato sauce with the garlic and tomatoes and a little more oil (see p196–7), and simmer until thick.

Combine all the ingredients, including the juice from the peppers, and warm through. Check the seasoning. Serve hot, warm, at room temperature, or cold. I like the middle two options best.

A ratatouille is even better with herbs: thyme, basil, bay and parsley are all good complements.

1 • A late merger. It's a vegetable stew: why don't you just cook all the ingredients together? Because they'd turn into a vegetable mush, that's why.

You'd fry the aubergines first, because they take longest to soften. After stirring them around for a bit in a pan from which they'd sucked all the oil, you'd add the onions and the garlic; then the peppers; then the courgettes; then the tomatoes. These vegetables will overcook and break up.

There isn't a short cut (apart from leaving out the sweating of the aubergines, and giving them a hassle-free bake). The ingredients of a ratatouille should retain their distinctive characters, but in the context of a binding of tomato and garlic. Combining the ingredients only for a brief warm-through, and letting them cool together, allows them as long to get to know each other (Fergus Henderson's phrase) as they need.

Potatoes

The general rule, as we have seen, is that if you value nutrition above culinary aesthetics you should cook vegetables quickly, in as little water as possible. But the advantages of boiling potatoes – whether slowly, to eat as they are, or rapidly, before roasting – are worth the loss of a few nutrients, in my opinion.

BOILED POTATOES

Peel and cut the potatoes into pieces of the size you want. Put them in cold, lightly salted water in a saucepan, and turn the heat under the pan to low/medium.[1] – SEE WHY YOU DO IT When the water reaches simmering point, regulate the heat under the pan to retain a gentle simmer until the potatoes are tender.

Scrape new potatoes to get rid of as much skin, or as many surface blemishes, as you like.[2] Cook them as above.

WHY YOU DO IT

1 • A gentle simmer. Floury, maincrop potatoes (Maris Piper, King Edward or Desirée, for example) often fall apart when boiled, because the outsides overcook before the heat penetrates the interiors. Steaming in particular causes uneven cooking: when potatoes are piled on top of each other, some get better access to the steam than others. Slowly bringing potatoes to a simmer helps to preserve their textures; their surfaces get less of a bashing if the water is maintained at a gentle simmer rather than at a rolling boil.

Salt speeds the softening. In theory, salted water should help to shorten the time between the softening of the exterior of the potato and of its interior.

A boiled King Edward is an unexciting thing, I know. But sometimes, with a creamy sauce or rich gravy, or with the juices from a roast chicken, it's just what you want. You can always make a mash on your plate with a fork and some butter.

2 • A boring scrape. Although new potatoes keep their textures better, they still benefit from this slow-boiling treatment. They have more tender skins than do maincrop potatoes; how much of the skin you scrape off is up to you. I'm pretty slack about it, because I find that scraping potatoes is a longer and more boring task than peeling them.

Making potato salad, you can boil the new potatoes in their skins, drain them, wait for them to cool a little, then peel off the skins, before cutting them up and tossing them in vinaigrette (see p37). Lord, I hate doing that: it takes me for ever.

MASHED POTATO

HOW TO MAKE IT

Method 1 (for perfectionists):[1 – SEE WHY YOU DO IT] heat a pan of unsalted water to 70°C. Peel the potatoes, slice them 20mm thick, and add them to the water, which you should return to 70°C and keep at that temperature for 40 minutes. Drain the potatoes, and put them in cold water for half an hour. Bring a pan of salted water to boiling point, add the potatoes and simmer until the point of a knife slips into them easily, then drain. Return them to the hot pan and stir them around, over a low heat if necessary, to dry. Push them back into the same pan – or into a clean one if you're fussy about the bits of potato that will have stuck to it – through a food mill or potato ricer. Add warm milk until the potato reaches the required consistency; add butter, and more salt if needed. Put the pan on to a low heat, above a heat disperser, and stir until warmed through. Or spoon the potatoes into a heatproof serving dish, cover loosely with foil, and put into a low oven (say, gas mark 1/140°C) for 10 minutes or so.

Method 2 (for the rest of us): peel[2] the potatoes[3] and cut them into slices about 1.5cm thick.[4] Put them into cold, salted water,[5] bring slowly to a simmer, and cook them at a gentle simmer until the point of a knife slips into them easily.[6] Drain, and follow the mashing instructions above.[7]

VARIATIONS

Simmer an unpeeled garlic clove – or 2 or 3 cloves if you like – with the potatoes, and put that through the mill too (it will leave behind its skin). Olive oil can substitute for butter; or, luxuriously, truffle oil.

Aligot – or, if you prefer, cheesy mash – is a delicious variation. Prepare the potatoes as above, with the garlic clove. Drain, mash, and add a melting cheese to the potatoes and butter in the pan (about 350g cheese to 900g potatoes); Cantal is authentic, because the dish comes from the Auvergne, but Cheddar, pecorino, Gruyère or anything similar is delicious too. Pepper and nutmeg would be good seasonings. Remember that the cheese is salty.

You could stir the mixture on a low heat, again with a heat disperser below the pan if possible, until the cheese is melted and the mixture is warmed through. That will release starch and produce a gluey consistency, which is often a feature of the authentic version. Or you could transfer it to an oven dish and warm it through at gas mark 5/190°C for 15 minutes or so. I sometimes leave it uncovered, browning the surface slightly.

Turning the dish into something that is not aligot, I like to add fried lardons, with their fat (instead of the butter). I then have a choice of simmering the garlic with the potatoes or frying it with the lardons; the flavour of the latter is more assertive. I also like it with red onions, softened gently in butter and a little oil. (Making this version, I don't add butter with the cheese.)

WHY YOU DO IT

1 • How far are you prepared to go to avoid glueyness in your mash? Method 1, with its two stages of cooking, is recommended by Heston Blumenthal and Jeffrey Steingarten. Some day, when my diary is empty, I might try it.

Glueyness comes from starch gel that has leached from broken-down cells. After 40 minutes in 70°C water, the cells expand; when you put the potatoes into cold water, the cells bond, and are less likely to rupture. Then you can cook and mash the potatoes safely.

2 • Peeling. Jeffrey Steingarten is reassuring on this point. It is not true, he says, that most of the nutrients are in the skin or immediately below it. There are plenty in the rest of the potato. That is welcome news, because it means that there is no good reason to follow those who tell you to cook the potatoes in their skins. How large are these cooks' saucepans, which can accommodate, say, half a dozen large

King Edwards and the water to cover them? Because the potatoes are big, their outsides cook some time before their interiors do. Then there's peeling the hot potatoes, a task that I am not dextrous enough to accomplish without scorching myself and getting little sticky bits of peel all over the place. By the time I've peeled six potatoes, they are almost cold; as you may have noticed, cold potatoes firm up. You have to bash them up harder, breaking the starch cells: result, gluey mash.

3 • What potatoes, and how many? Select a floury, maincrop variety, such as King Edward and Maris Piper. Some say that Desirée, which may be a little waxier than other maincrop varieties, do not mash so well, but I've had perfectly acceptable results with them.

The term 'waxy' is used to describe the firmer consistency of new potatoes. Joel Robuchon, creator of the most celebrated mashed potato in the world, uses a waxy variety. Waxy potatoes contain less starch; but I, lacking Robuchon's skill (and not wanting to use as much butter and cream as he does), have found that they produce a more gluey result, perhaps because mashing them requires greater force.

I use about 900g potatoes for 4 people, with a fairly indulgent 75g or so of butter.

4 • Slicing the potatoes. If you leave them whole, the outer part of the potatoes will be overcooked before the inner part has softened; the cells in the overcooked potato will be likely to break. If you cut up the potatoes into tiny pieces, you risk overcooking and waterlogging them; you'll also lose a good many nutrients to the cooking water. Slices should cook more evenly than would chunks.

5 • A cold water start. See Boiled potatoes, p176.

6 • When they're ready. The potatoes you're going to mash have to be soft, of course; but you need to catch them at the moment between their getting tender and falling apart. As I say above, starchy cells in overcooked potatoes will break, leaking gluey stuff into your mash.

7 • How to mash. Drain the potatoes, then return them to the hot pan for a minute, to allow some water to evaporate. Don't bash them about and break up starch cells. Then mash.

Delia Smith suggests you mash potatoes with a handheld, electric whisk. I am sure that Delia has never made gluey potatoes in her life, but I do not recommend her technique; again, it risks smashing the starch cells, releasing the gel. Certainly, you must never try putting the potatoes into a food processor. The blades will convert them into wallpaper paste.

A hand-held masher has the drawback of forcing you to mash portions of potato several times as you seek to obliterate all the lumps. Several experts recommend potato ricers; but I don't possess one. I use a food mill (p26), which produces a smooth, unsticky result. Joel Robuchon, apparently, forces his milled mash through a sieve.

Reheating mash needs care: while it will withstand a beating from a wooden spoon, it will turn pasty if you cook it again. Put it on to the lowest heat, with a heat disperser preferably, and stir it regularly and gently; warming it in the oven may be a safer option. Don't reheat for longer than 10 minutes.

ROAST POTATOES

HOW TO MAKE THEM

Turn on the oven to gas mark 6/200°C, put some fat[1] – SEE WHY YOU DO IT into a roasting pan and put the pan on to a high shelf. Bring a saucepan of water to the boil. Meanwhile, peel and cut up your potatoes[2] into pieces of whatever size and shape you like. Add them, and salt, to the saucepan, bring the water back to the boil, and simmer for 5 minutes.[3] Drain, put the potatoes back into the warm saucepan, and stir them over a low heat to dry them and rough up their edges.[4]

Take the roasting pan out of the oven and tip in the potatoes, turning them in the sizzling fat.[5] Keep the potatoes separate, so that they roast rather than steam. Put them back into the oven. Check them after 20 minutes; if they are browned underneath, turn them. (Using my oven, and my roasting pan, I can leave them for half an hour.) If you want to turn them again on to their remaining pale sides, do; they should be tender in 45 to 50 minutes.[6] Tilting the pan so that the superfluous fat falls away, remove the potatoes to a colander lined with kitchen paper. Add salt, toss a little, and serve.

VARIATIONS

Five minutes before the end of cooking, add some rosemary, stripped from its branches, and stir in. Put some garlic cloves to roast with the potatoes – but beware of cooking the garlic for longer than 30 minutes, or you might find that there's no

182   DON'T SWEAT THE AUBERGINE

pulp left. You could also take a whole bulb, rub a little olive oil over it, wrap it loosely in foil, and put it in the pan with the potatoes. Roast potatoes are delicious with the creamy garlic you squeeze from the husks.

Or don't parboil them. Simply peel them, rinse them in water, pat them dry (to avoid the violent reaction of water and hot fat, and to begin the browning process as soon as possible), and turn them in the hot fat as above. The point of parboiling – about which more below – is to dispel surface starch and to rough up the surfaces of the potatoes, enabling them to crisp. The surface of a non-parboiled roast potato may be slimy and chewy. And it will stick to all but the most efficient non-stick surfaces. Nevertheless, the potato will have a concentrated, earthy sweetness that a boiled one will have lost.

One of my favourite dishes is shoulder of lamb, slow-roasted above a bed of sliced potatoes, which have been mixed with garlic and rosemary and covered with water or stock (Lamb boulangère – see p222). When the lamb is ready, you remove it from the oven, turning up the heat to brown the potatoes and evaporate most of the liquid. The potatoes are deliciously imbued with fat and juices from the meat.

Do not parboil new potatoes before roasting them (see below).

WHY YOU DO IT

1 • What fat, and how much? Animal fats – beef dripping, lard, goose or duck fat – work best, some claim. I think that olive, sunflower or vegetable oil work fine. Use enough to provide a layer in the pan, and to coat all the potatoes with some to spare.

2 • What potatoes? I have assumed that you're using floury, maincrop potatoes – although roasts made with Desirées are, I have found, a little dry. You can roast new potatoes too, of course; but I don't agree with those who say that you should leave them in their skins, which, in my experience, go tough. So peel them or, in the case of Jerseys and the like, scrape them thoroughly; put them straight into the hot oil. Less starchy than maincrop varieties, they don't benefit from parboiling.

3 • Parboiling. Three procedures help to give you crunchy roast potatoes: parboiling them, drying them, and tipping them into hot fat.

I prefer parboiling to steaming. Steaming does not seem to get rid of all the surface starch, and the potatoes are more likely to stick to the roasting pan. Some writers insist that you simmer the potatoes until they are tender and nearly falling apart. I do not think that they need to spend more than 5 minutes in the simmering water. It's the surface of the potatoes that concerns you; the insides will cook in the oven, and will be just as fluffy. You don't need to start the potatoes from cold, and you can keep the water at a rolling boil if you like (see Boiled potatoes, p176). The point here is to cook and bash up the surfaces – processes that the salt in the water will hasten.

4 • Rough surfaces. Stirring the potatoes in the hot pan dries them, so that they do not repel the fat when they hit it; it also roughs up the surfaces, helping them to crisp. Some cooks dust the potatoes in flour to create a crispy surface; I don't think that's necessary.

5 • Hot fat. I have tried turning the potatoes in cold oil, direct from the bottle; I have also tried roasting cold potatoes. Both options work, but a hot, dry potato tipped into sizzling fat acquires the crunchiest surface.

6 • Ready in 50 minutes. Assuming they're on their own in a gas mark 6/200°C oven, that is. Of course, you are more than likely to be preparing them to accompany some other dish that is cooking there: a roasting meat or a simmering stew. It is likely, too, that you will be cooking these dishes at a lower temperature than gas mark 6/200°C.

If you are cooking a stew, your oven will be (I hope, for the stew's sake) at too low a temperature for the potatoes. Consider transferring the casserole dish to the smallest ring on the hob, and putting a heat disperser underneath it. Your roast may be cooking at gas mark 4/180°C; put it on to a low shelf, with the potatoes on a high one. You'll want to take the meat out of the oven at least 20 minutes before carving; at that point, you can turn up the temperature to get the potatoes good and crisp. If you're slow-roasting some lamb or belly pork, say, at a temperature of gas mark 1/140°C or less, try taking it out of the oven before you put the potatoes in; turn the temperature right up (to gas mark 8/230°C), in order to cook them in 30 to 40 minutes, but keep an eye on them. The meat will still be warm when the potatoes are done; I prefer it at this temperature.

FRIED POTATOES

You can follow the procedure for roast potatoes: peel, cut up, parboil, dry, then fry. You must fry them in a single layer, as the recipe books say. How large are these writers' frying pans? I've got a pretty large one, and I find it hard to cram it with enough potatoes to satisfy a family of four. Cooking them requires attention: you need to turn them and move them around, to make sure they cook evenly. Grumble, grumble. Still, they taste good.

CHIPS

I don't make chips often at home. I used to think that they were best left to restaurants, until I learned that even quite prestigious restaurants rely on oven chips. But I still think that maintaining a chip pan, filled with several bottles of rapidly degrading oil, is a bore for the home cook.

Nevertheless, you can get good chips with any old pan, and without a thermometer. It's quite satisfying.

The two-phase method of cooking chips was commonly accepted as the best, until Heston Blumenthal came up with his three-phase version. Like his mashed potato recipe, it's for perfectionists rather than for home cooks merely trying to get meals on to their tables.

The two phases are an initial one of about 10 to 15 minutes, at 140°C; and then one at 190°C, until the chips are brown. During phase one, you're boiling the chips in the oil; phase two is the browning bit. If you tried to combine these phases, the potatoes would brown before the interiors were cooked; also, the first phase – for reasons I don't fully understand, but something to do with what happens to the surface starch – helps to produce a crunchier chip.

HOW TO MAKE THEM

If you have a proper chip pan or a thermometer, this process will be easier. Half fill a large saucepan with oil (I use a mixture of corn and sunflower, or corn and groundnut). DO NOT GO ABOVE HALFWAY UP THE PAN WITH THE OIL. Even potatoes that you think you've dried thoroughly will cause the hot oil to erupt when they hit it; an overflowing pan may give you all sorts of trouble.

Peel and cut the potatoes: choose a thickness, cut them horizontally, then cut these rounds horizontally or vertically, depending on the size of chip you want. Put them into cold water, to remove some of the surface starch; dry them.

Heat the oil, above a low to medium flame. Drop in a small piece of bread: it should sizzle gently. Add the potatoes, and don't crowd the pan; you may need to cook them in several batches. Cook for 10 to 15 minutes, until nearly tender.

You're told to remove the potatoes at this stage. I never bother. Simply turn up the heat, and cook the chips until golden. Remove them to a colander lined with paper towels. If you're cooking them in batches, keep them warm in a low oven. Add salt just before serving.

GRATIN DAUPHINOIS

For 4

4 medium-sized potatoes[1] – SEE WHY YOU DO IT
1 pot (284ml) cream
Milk[2]
1/2 garlic clove, chopped[3]
Pinch of nutmeg
Salt
1 dstsp clarified butter or olive oil or vegetable oil[4]

Heat the oven to gas mark 3/160°C.[5] Slice the potatoes no thicker than a pound coin, and as thinly as you like, into a bowl of cold water. Drain them. Elizabeth David says that you should dry them at this point, but that seems unnecessarily fussy. Smear the clarified butter or oil over the insides of a gratin dish. (A Pyrex oven dish is fine.)

Put the cream and some milk (about 150ml to start with) into a saucepan of roughly the same diameter as the gratin dish; add the garlic, nutmeg and a little salt, and tip in the potatoes. Arrange them as economically as possible, and add more milk, if you need to, until it comes level with the top layer. Heat the pan gently. When it starts to bubble, tip the contents into the gratin dish, tidy up the potatoes, and bake for about 1 1/2 hours, or until the surface is browned and the potatoes are sitting in a reduced, wobbly liquid.

VARIATIONS

Another way to prepare gratin dauphinois is with cream only. Unless you use an artery-blocking quantity of the stuff, you'll probably find that it doesn't cover the potatoes; so, in order to cook the exposed potatoes (by steaming) and to prevent the cream from reducing and thickening too quickly, cover the gratin dish with foil. (Nevertheless, it's probably worth using a little more cream than in the main recipe – 350ml, say.) After about 50 minutes (at gas mark 3/160°C), check on progress: if the potatoes are tender but the cream is still runny, uncover the dish and cook for 10 minutes or so longer. If the potatoes are not tender, you may need to turn up the oven. If the potatoes are not tender and the cream has evaporated, er ... did you cover the dish tightly enough?

WHY YOU DO IT

1 • What potatoes, and how many? Waxy potatoes such as Charlottes work best, because they retain their textures. But any will do.

'Small portions please,' write Alastair Little and Richard Whittington in their recipe (*Keep It Simple*) for gratin dauphinois. 'I'll serve and eat as much of it as I like, thank you very much,' is my response to that; but I concede that a portion of dauphinois containing the equivalent of one medium-sized potato for each person is about right.

2 • How much liquid? One way of cooking gratin dauphinois is to layer the potatoes in a dish, and then pour over hot milk and cream; but it's hard to know how much milk and cream you'll need. That's a reason why I suggest putting the ingredients into a saucepan first: then you can add liquid until you have enough. A second reason is to enable you to heat the ingredients. It might take them half an hour to reach simmering point if placed cold in the oven.

3 • How strong do you want the garlic? You could flavour the dauphinois with a hint of garlic by rubbing a cut clove over the gratin dish, which, if it's earthenware, will absorb some garlic and give it back to the dauphinois during cooking. Or, for a stronger flavour, cut up the garlic and then crush it (using a mortar and pestle, or with the back of a strong knife on a chopping board) with some salt. Mix it with the butter or oil that you smear over the dish.

4 • Clarified butter. The milk solids in ordinary butter can cause food to stick. Smear on the clarified butter or the oil with your fingers, or with a paper towel (which of course will absorb a good deal of it). To clarify butter: melt unsalted butter in a saucepan over a very low heat. Pour the yellow butter into a cup or jar, leaving behind any foam or solid material. Cover, and refrigerate.

5 • Oven heat. It is impossible to give a recipe that you can guarantee will work every time. Containers will vary in heat conduction and surface areas; potatoes will vary in their propensities to absorb liquid. I can prepare a dauphinois one week and find it in perfect condition after an hour in the oven; the next week, I will follow a similar formula and find that at the same point the potatoes are still swimming in milk and cream.

Most recipes suggest that gas mark 5/190°C for 50 minutes to an hour will work; and, often, it will. I prefer to cook my dauphinois more slowly, at gas mark 3/160°C, for about an hour and a half. Check it after 30 minutes. Is it bubbling gently? Good. Check it after 45 minutes. Has it started to thicken? Not at all? Then turn up the oven. Keep checking. The dauphinois is ready when the surface has browned and the potatoes are sitting in a thick, reduced cream. The one issue you don't need to worry about is whether the potatoes will be tender: the milk and cream are excellent aids to cooking.

OTHER BAKED, SLICED POTATO DISHES

There's a variation on the cream-only version of gratin dauphinois in Rose Gray and Ruth Rogers' first River Café book. You fry cubes of pancetta (100g for 4, say) and garlic before adding them to the cream and potatoes, with sage as well.

Some gratin dauphinois recipes include egg and/or cheese. I wouldn't use egg, because I prefer, in this context, the consistency of thickened cream to that of custard; and achieving that custard is not straightforward – you're more likely to find bits of scrambled egg among your potatoes. But I do use cheese – Gruyère or Cantal – occasionally, in a cream-only gratin. The problem here is the acidity of the cheese, which can inhibit the softening of the potatoes. Add grated cheese towards the end of cooking, either stirring it in or spreading it on the surface, to provide a browned crust.

Sometimes I fry pancetta and garlic in a casserole, then add sliced Charlotte potatoes (900g) and a couple of sliced onions (for 4 people). I bake it in the oven (at gas mark 3/160°C again), giving it a stir every quarter of an hour to merge the onion and potato as the onion softens. The onion provides enough moisture to prevent the potatoes from sticking. The mixture emerges golden and melting. A richer version might include cream and/or cheese as well.

Gratin Savoyard is cooked in stock. You mix sliced potatoes, garlic, Gruyère cheese, salt, pepper and nutmeg, layer them in a dish, pour over hot stock to cover, and bake, uncovered, in a gas mark 5/190°C oven (for 4: 900g of potatoes and 120g of Gruyère). Again, your best bet may be to add the cheese at the end; the potatoes will take longer to cook than they do in milk or cream anyway, and the liquid will not reduce so quickly. You want a browned surface and just enough liquid to moisten each portion. Sliced onions are an optional, inauthentic variation; thyme would be a nice addition.

Potatoes boulangère is similar, but without the cheese.

Potatoes Anna are layered with butter. Grease a gratin dish. Peel the potatoes. Cut them into rounds, arranging them in the dish; when you have a layer, cut slivers of butter and place them on top, and grind over salt and pepper. Build the layers of potato, butter and seasoning. Cover the dish loosely with foil, and bake at gas mark 4/180 C for 45 minutes. Uncover the dish, and bake for another 30 minutes, turning up the heat if the top layer is not browning.

Root vegetables: celeriac, parsnip, turnip, swede

A celeriac is a pain to peel, and you can risk serious injury attempting to get a knife through a large swede. Try not to let these drawbacks put you off these underrated vegetables. Too often in winter we carry on buying out-of-season aubergines, courgettes and tomatoes, ignoring humbler, cheaper, tastier roots.

All may be cut up, tossed in oil, and baked. They go well with spices including cumin and coriander. Young turnips may be cooked as you would carrots (see p152), with a little stock or water and butter. Celeriac or parsnip, steamed apart until tender, make superb mash with an equal quantity of potato. Another delicious mash to go with roasts consists of equal quantities of turnip, swede and carrot; steam all three, taking care not to overcook the carrot; blitz in a food processor; and warm through with a little salt and nutmeg, as well as plenty of butter.

Spinach

The greater the quantity of a vegetable, the more difficult it is to cook it efficiently. Unfortunately, you need a huge bagful of spinach just to feed four people.

You could plunge it into a large vat of boiling water; some say that this is the best way, because it's the quickest. You cook it for a minute or two, drain it, plunge it into cold water, drain again and squeeze out the excess liquid, then warm it through with butter and seasoning. As I've said elsewhere, I'm not a fan of this method, which loses nutrients and, I think, flavour.

So: wash the spinach thoroughly; you may need to do it twice (wash the bagged supermarket spinach too). Shove the wet spinach into a pot, put on the lid, and put the pot on the hob above a maximum heat. After a minute or less the spinach will start to collapse; you can take off the lid and stir the leaves, getting them all into contact with the now boiling liquid they have shed. Continue to stir for another minute or two. The leaves won't take long to wilt and become tender. Drain, and push down on the spinach in the colander with a wooden spoon to get rid of as much moisture as possible. Return the spinach to a pan with a generous portion of butter; add salt and pepper (nutmeg too, if you like), and stir to coat the leaves and warm them through.

If you're going to cook the spinach and reheat it later, my advice is not to follow injunctions to refresh it in cold water, because that process always seems to annul the flavour; instead, shorten the initial cooking, to allow for the cooking that the spinach will continue to undergo before it cools down.

Tomatoes

The term 'salad tomatoes', as used by greengrocers, is almost always a lie. Those hard, orange things, acidic but otherwise tasteless, are not happy additions to a cold plate; and they're not worth cooking either. Plumper, vine tomatoes are a better bet, although the message that the vine is meant to convey – that the tomatoes have ripened in the sunshine – is probably also a con. They sometimes have the texture of soggy blotting paper, perhaps because they have been refrigerated; but, with a boost from other ingredients, they can form the basis of a reasonable sauce.

Deseeding tomatoes is one of the jobs that separates the professional cook from the rest of us. Most recipes involving tomatoes tell you to skin and deseed them. These writers have access to better tomatoes than the ones I get hold of: after I've deseeded one, there's not much left. They are also prepared to discard the acidity of the juice in favour of the sweetness of the flesh; but I like the balance between those tastes. You could scrape the juice and seeds into a sieve, returning the juice to the flesh; are you that fussy about tomato seeds?

Deseeding is worth doing if you don't want the juices to soak other ingredients: when you're making certain salads, for example, or sandwiches. Bring a pan of water to the boil, turn off the heat, and drop in the tomatoes for 20 seconds. Drain, and cool them with cold water. The skins should come away easily. Cut the tomatoes in half, and scrape out the seeds and juice.

However, I like Nigel Slater's throw-it-all-in method of making tomato sauce (from *Appetite*). I have adapted it slightly to include vinegar and sugar: even the best tomatoes benefit from an injection of extra acidity and sweetness. Tomato paste, to give a greater concentration of flavour, is optional: judge for yourself how much your fresh or canned tomatoes need this assistance.

Simmering the sauce thickens it, of course; it also mellows the flavour.

TOMATO SAUCE 1

For 2 to 3

Finely chop a clove of garlic, and soften it for a minute or
so in 1 tbsp olive oil in a heavy saucepan. Add 1 dstsp white
wine vinegar and 1 dstsp tomato paste, and cook this mixture
for a minute; then throw in 6 roughly chopped, medium-sized
tomatoes, a pinch or two of sugar, and a little salt. Cook gently
until the tomatoes have broken down and the sauce is gloopy:
maybe 30 minutes or more. The skins are not worth eating. You
can remove them by passing the sauce through a food mill
or a sieve. Check the seasoning. Add herbs if you like: some
torn-up basil, or some oregano.

TOMATO SAUCE 2

HOW TO MAKE IT

For 2 to 3

Follow the procedure for sauce 1, but with a 400g can of
tomatoes and their juice. Once they're simmering, you can
break them up with a wooden spoon or mash them with a
potato masher. If you want a smooth texture, put the cooked
sauce through the food mill.

TOMATO SAUCE 3

For 2 to 3

Put the roughly chopped tomatoes, or the can of tomatoes and their juice, into a saucepan, and simmer with a pinch of salt until thick. Pass the sauce through a food mill or a sieve if you like. Warm through again, take off the heat, and add 1 tbsp extra virgin olive oil and a clove of garlic, crushed with some salt. Add any other herbs and seasonings you like.

The fruity olive oil and the fresh garlic will of course be much more prominent in this sauce than in the other two. So you need good olive oil, and good garlic. You also need to be prepared to carry the evidence of what you've been eating on your breath.

VARIATIONS

Soften a chopped onion after you've softened the garlic. Add a whizzed dried chilli or two before adding the tomatoes; or add whole dried chillies to the sauce as it cooks, and remove them before serving. Add 1 tsp harissa (p134) to the sauce as it cooks.

Gently fry cubes of pancetta first, then add the garlic, and onion as well, if you want it.

For a puttanesca sauce for pasta, soften half a small tin of anchovies with the oil and garlic, add dried chilli if you like, and proceed with the tomato sauce as normal. When the sauce is nearly cooked, add a dozen stoned, roughly chopped black olives, as well as 25g capers, rinsed of their vinegary juice. Add chopped parsley too, if you like.

BAKED TOMATOES

Cut the tomatoes in half, horizontally (i.e., at right angles to the stem). Place on an oiled baking tray. Insert little slivers of garlic in the compartments holding the juice and the seeds, drizzle olive oil on top, and sprinkle with salt. I like to cook these tomatoes slowly, in a gas mark 1/140°C oven; it usually takes a couple of hours for their flavour to concentrate. But I can't give you a reason why cooking them at gas mark 6/200°C for 20 minutes to half an hour shouldn't serve just as well. You could cook some spaghetti, toss it in butter or oil, and serve these tomatoes on top. Or push the tomatoes through a food mill, add a little more olive oil, and toss the pasta with this purée.

Or: bake the tomatoes with just the salt, and serve as a warm salad with a vinaigrette (p37).

We may justify to ourselves the killing of animals for food, but we surely cannot feel comfortable about treating them with abominable cruelty. That's one argument for buying free range or organic meat; another reason, more self-interested, is that it tastes better. A third is that we can feel more confident about the provenance of what we're taking home. A fourth is that we can feel more confident that the meat hasn't been pumped full of substances we'd rather not ingest.

You can buy a battery chicken for £3. An organic one may cost four times that amount. The price of the battery chicken, the pathetic victim of abuse in the name of cheap food, seems to me to be obscene. That may be easy for me to say: I can afford to pay more. I make sure, though, that I get my money's worth from the chicken I buy. The meat from a 2kg bird will provide two meals for my family; from the carcass, I make a stock, and get a further meal from that. The £12 is good value.

Lean meat is very expensive. Less glamorous cuts, however, are reasonably priced, and are often more interesting. I'd rather eat the steak the French call onglet, and that we call skirt, than fillet. Shoulder of lamb can be more characterful than leg. Nothing is better than slow-roasted belly pork (p217).

A large bird or joint of meat going straight from fridge

to oven will take longer to cook than would one starting off at room temperature. As Richard Ehrlich has shown, instructions to take the meat out of the fridge an hour before cooking are useless: the rise in temperature in that time will be insignificant. To reduce the cooking time, you'll need to get your Christmas turkey out of the fridge at least four hours in advance. It's probably worth doing: the less time the tender breast meat spends in the oven, the better.

Roasts

A roast requires very little preparation, and would provide the most foolproof of all meals if only we had the luxury of a set of unvarying rules about cooking times.

It's not in the interests of cookery writers to advertise that their recipes are fallible. But even they will admit that roasting times and oven settings can be only rough guidelines. You have to keep an eye on the meat, particularly if you're fussy about its level of doneness. If nothing much is happening, turn up the dial – but be wary of drying out the meat by roasting it at a high temperature for too long. If the meat is sizzling too angrily, turn down the dial. You'll learn how to tell when a joint is ready through experience; or, if you want to be punctilious, you could invest in a meat thermometer. If the centre of a piece of lamb or beef is 60°C, the meat is medium done; pork is medium done at 70°C.

The convention is to baste a joint at regular intervals. It may not be necessary. A chicken covered in skin, or a pork joint covered in crackling, or a lamb joint covered in fat, is not going to get extra benefit from having juices poured over it. A lean piece of beef might.

ROAST CHICKEN

HOW TO MAKE IT

Wipe the chicken inside and out with paper towels.[1] – SEE WHY
YOU DO IT Mix 80g to 100g of butter with salt, pepper if you like, a
chopped garlic clove and herbs – thyme, tarragon or rosemary
work well. Rub the butter all over the chicken; you can put
some inside the skin covering the breast, too, as well as in the
cavity. Put the chicken in a roasting pan in a gas mark 6/200°C
oven. After 25 minutes, pour a glass of wine into the pan (not
over the breast of the bird), and turn down the oven to gas
mark 3/160°C.[2] Cook the chicken for a total of 20 minutes
for each 500g it weighs, plus a further 30 minutes. (A 1.5kg
chicken, therefore, will take one hour and 30 minutes.[3])

Take the chicken out of the oven, and pour the liquid into
a saucepan. Leave the chicken to rest in its pan for at least 15
minutes.[4] Just before you are ready to serve, heat up the sauce,
and check for seasoning. The chicken will have exuded some
more juices; pour them into the rest of the sauce. Either pour
the sauce over the chicken or serve it in a separate jug.[5]

VARIATIONS

I got the above method from Hugh Fearnley-Whittingstall's *River
Cottage Meat Book*, and I love it. I think that it's the recipe that
best conforms to the ideal of eating a good chicken: simple,
because you don't have to do any fiddling with the sauce,
yet also luxurious, because of all that butter. It wants only the

simplest of accompaniments: some boiled potatoes or rice, with a salad to follow. The chicken and its sauce are, quite properly, the centre of attention.

You could put the **giblets** (but not the liver, which, as I said when discussing stocks, makes sauces bitter) into the roasting pan at the start, as well as a chopped onion and carrot, some unpeeled garlic cloves, and perhaps a few extra chicken wings – all to add flavour to the sauce.

Stuff the chicken with unpeeled **garlic cloves**, and eat the meat with the pulp, squeezed from their husks. Be aware, though, that it's quite hard to predict how much cooking a garlic clove will need before the garlic achieves that soft, sweet quality. You could put a whole head of garlic, cut in half, with the chicken, and at the end push out the pulp into the sauce with a wooden spoon.

Lemon goes very well with chicken. Put a whole lemon inside the cavity, or lemon halves or quarters; or do that, but squeeze the juice over the chicken first.

Or: don't pour the **wine** into the pan while the chicken is cooking. (But check that the buttery juices aren't in danger of burning on your roasting pan; if they are, add just enough wine, stock or water to keep them liquid.) Wait until the end, remove the chicken from the pan, tip the juices into a saucepan, put the pan on to a medium heat on the hob, and throw in your wine. The advantage of this method is that there will be tasty, caramelized residues in the pan to incorporate into your sauce. Stir them in, bubble the wine to evaporate its alcohol and to lose some of its acidity, and add this sauce to the other juices in the saucepan.

This is a gravy – one with a high fat content. But gravies are more commonly prepared as below.

I am not convinced that **pot-roasting** – cooking the chicken, either browned first or at the end, and served with the sauce that it and the vegetables in the pot produce – is worthwhile. If you want to cook your chicken in a pot, in my view, you might as well joint it first, add the breast portions just 25 minutes or so before the end, and call it a stew. But pot-roasted, fatty cuts of pork or lamb can be good. See below.

GRAVY

Of course, there are many other ways of preparing a chicken. You might want to accompany it with a proper gravy. Rub a walnut-sized knob of butter, or the equivalent quantity of olive oil, over the chicken before you cook it, and season the bird. At the end of cooking, remove the chicken to a warm plate. Pour the juices in the pan into a bowl. Put the pan on to a ring of your hob, turn the heat to medium, and add some liquid – it might be water, or chicken stock, or a small glass of white wine, or a small glass of cider, or a tablespoon of vinegar. Bubble the liquid, stirring and scraping any residues from the pan into it. (This process is 'deglazing' – see p55.) Pour this sauce into another bowl.

Do you want to thicken your gravy? If you do, take a dessertspoon of the fat that will have risen to the top of the first bowl (the one into which you poured the first lot of pan juices), and put it in a small saucepan.[6] Carefully spoon the rest of the fat into the bin, or into a saucer for scraping into the bin later.[7] Add a dessertspoon of flour – or enough to make a roux the consistency of wet sand – to the fat in the saucepan, and heat gently until the mixture browns a little. Then turn up the heat to medium. You have two bowls, one with the degreased pan juices, the second with the liquid you produced by deglazing the roasting pan. Pour their contents, little by little, into the roux. You add a bit, stir vigorously until it's incorporated, then repeat the process until you get the consistency you want. It's just like making a béchamel (p46).

I suggest that you want a gravy of about the consistency of single cream; or perhaps, since you need to bear in mind that it will thicken as it cools, slightly thinner. Having used 1 dstsp flour, you'll probably need 150ml to 200ml of liquid to reach that consistency – that should be enough sauce for 4 to 6 people. So you may need some more stock. Don't add more alcohol at this stage – it will taste raw. If you don't have enough stock, revise your expectations downwards, and use just a teaspoon of flour, which should cope with the liquid in the bowls.

You could make a little stock while the chicken is cooking. One way, as in the recipe on p203, would be to put giblets and vegetables, along with wine or water or both, with the chicken in the roasting pan; the disadvantage of this method is that the bottom of the pan won't give you any tasty, crusty residues. Or you could put the giblets (if you're lucky enough to have them) with some vegetables in a pan and make a simple stock (see p51). If you don't have giblets, buy a few wings; or chop the wings from your chicken and use them. Or you could chop off just the wing tips, using them to add some chickeny flavour to a vegetable stock (p58).

By this time, you might be starting to see why I have come to like the version on p203 so much. And there is more work to do, if you want your gravy to taste as refined as possible. Put the pan on one side of a ring, and continue to simmer very gently. On the cooler side of the gravy, away from the heat, you'll get some skin, which you can lift off. You can continue with this process until your impatience overcomes your fussiness. Check the seasoning, and serve, in a warm sauce boat.

I used to like getting plenty of thick, dark gravy to cover my food. It seemed to me to be unsatisfactory to have meat and veg sitting on a thin puddle. But my taste has changed: I've come to recognize that the disadvantage of thickeners is that they inhibit flavour, and that thin sauces may be more satisfying for the taste buds. You can follow the procedure above without the flour stage: throw away as much fat from the meat juices as possible, then combine these juices with the sauce you made by deglazing the pan, and add as much stock as you need to bring your gravy to the required quantity. You could skim this gravy, as above, too.

ACCOMPANIMENTS

Roast chicken with gravy can take more elaborate accompaniments than the version on p203. Much as I love roast potatoes, I've come to feel, for some reason, that they don't go particularly well with roast chicken; I'd prefer a gratin dauphinois (p188). As the chicken and the gratin are both pale, serve a green vegetable with them; or carrots, perhaps.

Cauliflower cheese (p153) goes well with roast chicken. Again, you've got two pale preparations; so, instead of serving potatoes as well, offer another vegetable.

WHY YOU DO IT

1 • You don't have to wash the chicken.You may want to dry it, and to get rid of any gubbins inside the bird; but bacteria there will be killed by the high heat of the oven. It won't be killed by cold water. I am indebted to Richard Ehrlich for the information that the Food Standards Agency advises against washing chickens, warning that the procedure is more likely to spread bacteria than to kill them.

2 • Oven temperature. Roasts benefit from an initial cooking at a high temperature; 25 minutes at 200°C will start off the browning reactions, also known as Maillard reactions, that meat-eaters find most toothsome. (Don't ask me to explain them: even Harold McGee calls them 'exceedingly complex'.) In the case of chicken, what you're browning is mostly the skin; but that, too, is desirable.

Recipes sometimes instruct you to brown the meat in oil before putting it in the oven. It's a good idea to give small roasts, such as poussins, an initial browning, because the oven alone won't get very far with the process in the short time they take to cook.

The browning, once started, will continue at the lower temperature, which is perfectly adequate to cook the meat. Maintaining the high temperature might cause it to dry out. Your chicken will probably cook perfectly well, after the initial 200°C blast, at gas mark 2/150°C, but, as I've said before, ovens may vary.

3 • Cooking time.You risk poisoning yourself if you eat chicken that is still bloody. Test by inserting a thin knife or skewer into the meat at the thickest part of the thigh: if the juices run clear, the chicken is safe to eat; if there's any trace of red, you need to cook the bird for longer.

There are recipes that tell you to cook a 1.5kg chicken at 200°C for 50 minutes to an hour. I have found that the juices are clear after that time, but that the leg meat is still a little unyielding for my taste. I prefer the formula I've given above (20 minutes for each 500g, plus 30 minutes); by that time, the meat is close to falling off the bone.

There is a problem with my timings, though. The breast meat certainly does not take that long to cook; and the longer it spends in the oven after it has passed its point of readiness, the drier and tougher it gets.

This is the challenge in all meat cookery. You are heating muscle fibres, which become dry and tough if they are cooked for very long; you are also heating connective tissue, which needs long cooking to become tender. Fat is also part of the mix; it keeps the meat lubricated.

The connective tissue is what harnesses the muscles. The bits of an animal that get most exercise have the most connective tissue; in a raw or undercooked state, they are tough. The tender cuts, such as fillet steak, have little connective tissue.

When roasting a chicken, you're cooking tender breast meat and tough leg meat at the same time: it's as if you were cooking fillet steak alongside chuck. You wouldn't cook fillet steak for an hour and a half; and that timing is not ideal for chicken breast, either. I have found, nevertheless, that the method on p203 produces perfectly acceptable results. Perhaps the butter helps to keep the breast meat tender; certainly, fat in 'marbled' cuts of meat performs this function.

I don't believe that the sauce in the roasting pan helps to maintain tenderness. Some writers seem to think that surrounding meat with moisture will keep the meat itself moist; but anyone who has eaten an unsuccessful stew will know that it's perfectly possible for meat to emerge from liquid in a dried-out state. The wine or stock in the pan will lower the oven temperature; but that effect will be cancelled out by the qualities of the steam, which is a more efficient heating medium than air. Liquid will 'actually increase the fluid loss', Harold McGee writes. The net effect in the basic recipe above – lower oven temperature, but increased heating efficiency – is probably neutral.

You might come across recipes suggesting you wrap the chicken in foil, uncovering it half an hour before it's due to be ready in order to brown the skin. As I say, steaming the chicken will cook it pretty efficiently: even if the temperature inside the foil parcel is lower than that of the oven, the breast meat will be dry after an hour or so – and you still have to subject it to a 30-minute blast of unmediated heat.

There is a solution that Gordon Ramsay and other chefs prefer: you separate legs and wings from the breast, and start cooking them first. I don't bother to do this when I'm roasting a chicken, but I do poach a chicken in stages.

4 • Resting. On removal from the oven, a roast has a lot of juice close to the surface. Start carving, and the juice will spurt out. As the meat rests, the hotter, outer meat transfers its heat to the centre, continuing to cook it and transferring juices to it as well; also, the meat fibres, as they cool, retain their moisture better. So a rested chicken, or joint of beef or whatever, will be a moister one. Some juices will leak out, even so; you can add those to your gravy.

I've already commented on writers' annoying habit of telling you to keep things 'in a warm place'. I've got a grill, above my oven, with a door that opens frontwards, providing a table. The best I can do is put the resting meat on that; with the oven on, it's warmish. (I warm plates and serving dishes inside the grill – not turned on, of course.)

However, one of the reasons why cooking a roast meal can be so stressful is that we get far too worried about serving every component of it piping hot. A rested roast will retain its heat for some time, but it does not have to be hot to be enjoyable. A lukewarm gravy would be disappointing, though.

5 • Carving. If your carving technique is as crude as mine, you may prefer to cut up the chicken in the kitchen, put the bits in a serving dish, and then either pour the sauce over them or keep the sauce in a separate jug. But a whole roasted chicken is an inspiring sight; it seems a shame not to show it off to your guests. So, in spite of the humiliation, I prefer to let them watch me wrestle with it. If the sauce is in the same dish, you risk splashing it about; so I usually keep it separate.

6 • Gravy in a saucepan. You could, as so many recipes instruct, make the gravy in the roasting pan. But it's tricky to work a roux, and then to stir in liquid and keep it lump-free, over such a wide surface. A small saucepan makes the job much easier; and it also means that the gravy, if simmered for a while, does not evaporate so fast.

7 • De-greasing. After you've got rid of a spoonful or two of fat, you'll find it hard to get at the rest without spooning up some of the precious juice as well. Try dabbing bits of paper towel on the surface; it will soak up the fat but leave the juice behind. Don't pour the fat down the sink: it will solidify and block your drain.

Here is something you can do with the **liver**, if you got one with the giblets. Scrape it free of stringy material; don't worry if you turn it into mush. Bubble 2 tbsp brandy or calvados in a saucepan until it has reduced to 1 dstsp. Chop half a garlic clove and sprinkle with a little salt; reduce it to a pulp with the back of a knife. Melt a walnut-sized knob of butter over a gentle heat in a saucepan; add the garlic, and cook for 30 seconds. Turn up the heat a little, throw in the liver, and stir it around in the butter; it should cook through – going grey – in a minute. Take the pan off the heat, and stir in the brandy or calvados, with more salt if needed. Grind over black pepper, or add a pinch of cayenne. Spoon this pâté into a bowl, mash it up with a fork, and put it in the fridge, where it will set. Spread the pâté thinly on toasted rounds of baguette; there should be enough for four, to go with drinks before the meal.

ROAST DUCK

Duck is tricky to get right. I'm sorry to be pusillanimous, but I'm going to give you three options, and leave it to you to decide which method best suits your taste in cooking or, if you're adventurous enough to try all three, which works best for you.

In all three cases, you should first prick the skin of the duck all over with a needle – you use such a fine implement because you don't want to penetrate to the flesh, causing it to release its juices. And, in all cases, keep the fat for roasting potatoes; if it pours into a roasting pan, spoon it out every so often; if it joins liquid you've been using for poaching and steaming, put the liquid in the fridge and spoon off the fat when it's settled on the surface.

Method 1: use the method and timings for roast chicken, above. The books say that duck takes less time to cook: that is not my experience.

Method 2: put the duck, breast side down, on a rack in a roasting tin, and cook it in the centre of your oven for 3 hours at gas mark 1/4 or 110°C. Turn the duck on to its back, and cook it for a further half an hour at gas mark 7/220°C, or until the skin is crisp. You have to hope that this blast of heat will not dry up the breast meat too much; it may be protected by the fat that remains. But you don't want too much fat still to be there, because its absence will help you to get crispy skin.

Method 3: I don't have the energy fully to explain this version, let alone to follow it very often. Still, it does produce good results. For the complete account see *The Perfect* ... by Richard Ehrlich (who reports version 2 as well, only with an initial temperature of 120°C – it's possible that the gas mark 1/4 setting works for me because I have a hot oven), who has adapted his recipe from Julia Child's *The Way To Cook*. First you steam the duck, over water and on the hob, for 30 minutes; steaming helps to render a good deal of fat. Then, to continue the cooking, you braise the duck, breast down, in the steaming liquid plus more and with some vegetables, in a covered pan and in a gas mark 5/190°C oven, for a further 30 minutes. Then you turn up the oven by one setting to gas mark 6/200°C, uncover the pan, turn over the duck, and roast it for 20 minutes or so, until the skin goes crispy. You may find that the skin crisps more easily with the braising liquid, from which you can make a sauce, removed from the oven.

ROAST TURKEY

A cooked turkey with a moist breast is one of the holy grails of cooking. It's hard enough to produce a tender chicken breast at the end of an hour and a half in the oven; a turkey may be in there for 3 hours or longer. You see all sorts of suggestions: fast cooking, slow cooking, brining, covering the breast or whole bird in foil (see the remarks about roast chicken and foil, above), covering the breast in buttered muslin. The technique I like best, probably because it's the least fussy, is one recommended by Nigella Lawson (*How To Eat*): you roast the bird breast-side down, turning it over only for the last half-hour of cooking. That way, the fat in the turkey's back percolates down towards the breast, where it's most needed. Use the temperatures recommended above for roast chicken; but, as turkeys increase in size, the average time it takes to cook each 500g decreases. Lawson gives 2 hours for a 4.5kg turkey, and 3 1/2 hours for a 9kg one. But do check that the juices from the thickest part of the thigh run clear.

OTHER POULTRY AND GAME

The smaller birds need gentle browning in a frying pan first, because they won't have very long in the oven. After browning, a poussin will cook in 20 minutes to half an hour; a pigeon in about the same time; a pheasant will take about 40 minutes.

ROAST PORK: LOIN

HOW TO MAKE IT

Get your butcher to score the rind; or do it yourself, with a Stanley knife, cutting vertical lines about a finger's-width apart. Pat the rind dry, and rub a little sunflower or vegetable oil into it.[1] – SEE WHY YOU DO IT Season the meat, and put it in a roasting pan into a gas mark 7/220°C oven for 25 minutes,[2] then turn down the oven to gas mark 3/160°C.[3] Allow 25 minutes for each 500g of pork, with an extra 25 minutes. Rest the meat for at least 20 minutes before carving.

WHY YOU DO IT

1 • Crackling. I must admit that I do not produce perfect crackling every time. I think the principles I follow are sound, though, if not foolproof.

The scoring of the rind facilitates dehydration, and allows fat from below to bubble up, crisping the surface. You do not need at this stage to encourage further dehydration by salting the rind: the heat of the oven will do this job thoroughly on its own. Anyway, you need some water to stay there, because it will be the medium in which collagen, the tough protein that gives the rind its rubberiness, dissolves. The high initial heat is supposed to perform this process, as well as to brown the meat. After it, you can salt the rind, if you like – you want to get rid of the water now, retaining only the skin's crunchy qualities.

If your rind is not crackling when the joint is ready, try removing it and frying it. Put it in a heavy pan, with a little oil, over a gentle heat, and turn it often, because it will burn easily. It is easier to control the heat by frying than it is by grilling or putting the rind back into a hot oven.

I have not had good results when I've tried to kick-start the crackling process by frying or grilling the skin first. The skin has browned before the collagen has dissolved, I think.

If you braise the pork, you do not have to forgo the crackling. Indeed, you may get better results, because of the efficient breaking down of the collagen by the liquid. At the end of cooking, slice off the rind, and cut it into smaller pieces if you'll find them easier to manipulate. Put them into a heavy frying pan with a little oil over a low heat, and turn them frequently to prevent burning. Their residual liquid content will, before it vaporizes, cause them to crackle loudly, and even to jump in the pan. You should produce deliciously crisp crackling in 10 to 15 minutes.

2 • Searing. Unless it is a small joint, without much fatty protection, pork will take a higher initial heat than will chicken.

3 • Cooking. The lean meat from the loin of pork – that includes chops – is hard to cook well. Pork, like chicken, has to be cooked thoroughly before it's safe to eat: and again, as with the lean meat of chicken, there is only a brief interval between the moment when it is ready and the moment when it dries out.

ROAST PORK: SHOULDER, SPARE RIB ROAST, BELLY

HOW TO MAKE IT

1 – SEE WHY YOU DO IT Prepare the joint and the rind as with the loin (p215), but this time put the joint in its roasting tin on to the floor of the oven at its lowest temperature setting – if you like, you can rest the meat on a bed of sliced onions, which you discard at the end – for about 5 hours (for a 1.5kg piece of pork belly), or longer.[2] Hugh Fearnley-Whittingstall says that an 8kg shoulder will tolerate 24 hours of cooking. At the end, slice off the rind, and fry it to produce crackling as in the loin recipe, above.

VARIATIONS

These are forgiving cuts, with plenty of fat and connective tissue; I like to cook them slowly until the meat is exceptionally soft. But you could use the higher temperatures and faster timings given above for loin joints, and produce perfectly nice results.

Pork goes well with oriental flavourings. For his 5kg to

8kg shoulder, Hugh Fearnley-Whittingstall gives the following paste: 1 tbsp of a spice mix consisting of 2 star anise, 2 tsp fennel seeds, 1/2 cinnamon stick, 4 cloves, and 1 tsp black peppercorns, all ground; stirred into a mixture of 5 garlic cloves (crushed and made into a paste), a 5cm piece of ginger (also made into a paste), 2 tsp dried chilli flakes, 2 tsp ground ginger, 1 tbsp brown sugar, 1/2 tbsp salt, 1 tbsp sunflower or groundnut oil, and 1 tbsp soy sauce. You're most likely to have a smaller joint than the one in his recipe, so you can revise these quantities accordingly – and anyway, it's the kind of list you can adapt to your own taste.

Or try pot-roasting.[3] Oil a heavy casserole dish such as a Le Creuset, lay a bed of sliced onions on the bottom, lightly oil your joint and season it to taste, cover, and put in a very low oven[4] for about 4 hours, or until entirely tender. You may find that the meat and onions throw off quite a lot of liquid, which you can skim, reduce in a small saucepan until flavoursome, and serve as a sauce. Again, if you want crackling, slice off the rind and fry it.

You're often told that salt draws out the moisture from meat. But brining – immersing in a salt solution – can make tough cuts of meat juicier and more tender. Fergus Henderson has a lovely recipe for brined and roasted pork belly (from *Nose to Tail Eating*).

Henderson's brine includes 4 litres of water, 400g caster sugar, 600g sea salt, 12 juniper berries, 12 cloves, 12 black peppercorns, and 3 bay leaves: you put the ingredients in a pot and bring to the boil, in order to get the sugar and the salt to dissolve. Allow to cool. Put the meat into a china, glass or other non-reactive container, cover with brine, put a lid on, then leave it in a cool place for 3 days (your fridge may not be big enough; I use my cellar). Take the meat out; dry the skin and treat it as above; roast in a medium to hot oven (I should guess Henderson means gas mark 6/200°C), allowing 1 1/2 to 2 hours for a 2kg joint.

Or you could roast the joint at the very low setting recommended above. If you're serving it for lunch, though, you might not want to get up at 7 a.m. to put it in the oven.

ACCOMPANIMENTS

Mashed potato (p177) is perfect with pork, I think. And green vegetables: particularly cabbage, greens or spinach. Pork doesn't give you much help in making a gravy, so I prefer to eat it with mustard. For apple sauce: peel, core and slice up apples, putting them into water acidified with lemon juice to prevent discoloration; drain and put them into a pan above a heat disperser on a low heat, and cover; as they cook, judge whether you need to add more liquid (try a little orange juice); cook until you have a rough purée, and add caster sugar to taste. It's up to you whether you serve it hot; it doesn't have to be.

WHY YOU DO IT

1 • **Spare rib and spare ribs.** A spare rib cut, from which your butcher might sell spare rib chops, is not the same as the spare ribs commonly used for barbecuing. The former comes from the shoulder; the latter are trimmed from the belly.

2 • **How low is low?** Your lowest oven temperature may, like mine, be 130°C, or more. And if yours is a fan oven, your food will not in theory be any cooler on the floor of it than anywhere

else. You may, in these cases, produce better results if you cover the roasting pan with foil. See point 3.

3 • Pot-roasting. Recipes for pot roasts may tell you that the moist atmosphere of a covered pot will keep your joint more tender than would the raw heat of your oven. This claim is not necessarily true. Water, as I explained above when discussing chicken, is a more efficient heating medium than air. So your result depends on the variability of the temperature of your oven. If you can set the dial at 100°C or lower, you can cook your joint more gently uncovered than in a covered casserole, where the temperature will be 100°C or – if there is a build-up of steam – slightly higher. But my oven's lowest temperature is 130°C, so I usually get more tender results from the casserole.

If your joint is too large for your casserole dish, use a roasting tin, and cover it with foil.

You do not need to brown the joint first. You'll find that the exposed meat and skin in the dish will gradually brown on their own.

4 • Getting the right temperature. I have two Le Creuset casserole dishes, which behave differently. The contents of the larger, oval one may take 45 to 60 minutes to reach simmering point in a medium oven – at least 15 minutes longer than they would need in the smaller dish. If you're cooking your pot roast to a deadline, you may want to start it off at a high temperature (gas mark 6/200°C, or higher), or you could find that at least a third of the intended roasting time has been a non-event. Once such a heavy dish is properly heated, however, it will carry on doing its job at the lowest oven setting. Gently does it is the best approach.

ROAST LAMB: LEG OR SHOULDER

HOW TO MAKE IT

Set the oven at gas mark 8/230°C.[1] – SEE WHY YOU DO IT Strip as much rosemary or thyme as you like from their stems, and add enough olive oil to make a spreadable marinade. Add salt and pepper, and massage the mixture over your meat.[2] Chop a head of garlic in half, and put the halves under the meat in a roasting pan. Put in the oven; after 25 minutes, turn down the temperature to gas mark 3/160°C. Cooking time: the initial 25 minutes plus 15 minutes for each 500g meat.[3] When it's done to your liking (some like it pink; others, grey), remove the meat to a warm plate and leave to rest for at least 20 minutes before carving. Tip the pan juices into a bowl; skim off the fat.[4] Deglaze the roasting pan with wine, vinegar or stock (see under Roast chicken, p203); warm through this sauce with the pan juices to accompany the meat.

VARIATIONS

If you want more, and thicker, gravy, follow the instructions under Roast chicken above.

How about **caper sauce**? Nigella Lawson's recipe (from *How To Eat*) is a kind of béchamel (see p46) with two parts milk to one part lamb stock (chicken stock would be acceptable); you add capers from a jar, without rinsing them of their vinegar.

Lamb boulangère (again, shoulder or leg) is cooked above sliced potatoes. The lamb is often boned; I'm not sure what benefit that brings. Smear the base and sides of a roasting pan with a walnut-sized knob of butter. Peel and finely slice potatoes (maincrop or new) into a bowl of cold water (to prevent discoloration and to rinse the surface starch); drain and mix with salt and finely chopped garlic, as well as (optional) thyme or rosemary and chopped onion; tip into the roasting pan, and put the meat on top. Add stock or water; it doesn't need to cover the potatoes, but might come halfway up them. Put in the oven at a low heat (gas mark 2/150°C) for 2 hours, or until the lamb is ready; remove the lamb to a warm plate to rest, and, in the 20 minutes that it is doing so, cook the potatoes at gas mark 7/220°C to brown them.

Shoulder of lamb, with its high fat content, will respond particularly well to slow-roasting or to pot-roasting, the techniques for which I describe in the pork section, above. Lamb shanks, too, are delicious when prepared in this way. When slow-roasting, you do not need to subject the joint to an initial high heat: the lamb will brown gradually as it cooks.

When pot-roasting, you might throw into the pot, along with seasoning, a bay leaf, some rosemary, an onion or two, and a head of garlic. You may be surprised by how much sauce you get. Strain it into a jug or bowl, let it rest for a while, and spoon off – into the bin, not down the sink – the fat that has risen to the surface.[4] Pour the sauce into a saucepan, and bring it to a simmer; allow it to reduce uncovered for a while if you think it needs a more concentrated flavour. Be mindful that the salt content, as a percentage of the sauce, will increase as the volume reduces. If you like garlic, you could squeeze into the sauce the flesh from the cooked garlic cloves. The meat, meanwhile, can stay warm in the pot, with the lid on.

ACCOMPANIMENTS

Lamb and gratin dauphinois (p188) are a dream combination. Roast potatoes are almost as good. Lamb goes very well with white beans or chickpeas; if you like, you could mash the chickpeas with some garlic, salt and oil to make a kind of hummus. Spinach is another happy partner.

WHY YOU DO IT

1 • **Searing**. For the point of this high heat, see Roast chicken, p203. Lamb, which has more fat than do chicken and pork, can take a 25-minute blast of 230°C without losing too much of this lubrication.

2 • **Don't mess about with the meat**. I have got fed up with recipes that invite you to make incisions in the lamb and stick slivers of garlic and rosemary, and sometimes anchovy, into them. The garlic sticks to my fingers, and then the rosemary does too; I spend ages faffing around, irritatedly conscious that more dextrous cooks can probably perform this task in less than half the time I'm taking. Never again. I think I have a culinary justification for giving up preparing lamb in this way: moisture seeps out of the incisions. So the flavourings go on the surface.

The advice about the cut bulb of garlic comes from Nigel Slater (*Appetite*). You can squeeze some of the pulp with a wooden spoon into the sauce. Instead of, or in addition to, garlic, you could put slices of onion under the meat.

3 • Timings. The books tell you that these timings and temperatures should produce lamb cooked to medium. For pink meat, allow 12 minutes for each 500g; for well-done, 20 minutes. Again: you cannot guarantee that your oven or joint of meat will conform to the theories. Keep checking.

4 • Skim the fat. Lamb has a good deal of it. You probably don't want a fatty sauce if you're serving the meat with a rich accompaniment such as gratin dauphinois, but may not be so fussy if you've got boiled potatoes or rice. As a home cook, you have the advantage over the restaurant chef – who would rigorously skim such a sauce – in not being obliged to sacrifice flavour for aesthetic standards.

ROAST BEEF

HOW TO MAKE IT

My best advice, for all cuts, is to follow the timings and temperatures I gave for lamb (p221). If you keep the temperature high throughout, as some recipes advise, you risk expelling the fat that gives the meat a tender feel in the mouth. Wipe the meat dry with paper towels, then rub a little oil over it (the oily layer cuts down the evaporation of liquid, but if you were to rub it on to a wet piece of meat it would be less effective), and some salt and pepper, even though they will hardly penetrate.

Beef will be all the better for a rest of half an hour after cooking, and that delay has the advantage of giving you time to bake a Yorkshire pudding. If the batter occupies the top shelf of the oven, though, getting the roast potatoes crisp will be a challenge – unless you have two ovens. Perhaps you could try this: start the potatoes on the top shelf, above the beef; when you take out the beef, turn up the oven to gas mark 7/220°C, put the potatoes on to the middle shelf and the pudding pan with its fat on top; add the batter when the fat is hot (see below). There's a good chance that the potatoes will crisp up.

YORKSHIRE PUDDING

HOW TO MAKE IT

For 4 to 6
280ml milk[1] – SEE WHY YOU DO IT
1 egg
A little salt
112g self-raising flour

Put milk, egg and salt into a bowl. Add flour, a tablespoon at a time, whisking gently with a balloon whisk as you do so.[2] The batter should have the consistency of single cream. Allow to rest for at least half an hour before using.[3]

Heat the oven to gas mark 7/220°C, and put a layer

of dripping (appropriate if the pudding is to accompany roast beef), lard or sunflower oil into a roasting pan; put the pan into the oven for at least 5 minutes, to get the fat hot.[4] Pour in the batter, and cook for about half an hour, or until brown, puffy and set. This should be enough for 4 to 6 people.

WHY YOU DO IT

1 • The ingredients. The quantities I've given used to be the standard ones. But they've gone right out of fashion. Now you see recipes involving 2 or even 3 eggs, or 2 eggs plus 2 yolks, or 2 eggs plus 2 whites. I think that these batters can be a bit rich, and I prefer, for Yorkshire pudding and for pancakes, the original version. Self-raising flour will, obviously, help the batter to rise; for pancakes, use plain flour. For a lighter batter, you could use half milk, half water.

2 • Gentle mixing. Recipes often tell you to make a well in the flour, and break the egg into it. They don't often give the follow-up instruction that is the point of the procedure: you add a little milk to the egg, then stir in the nearby flour, then add some more milk, gradually working your way in a larger circle around the flour so that you stir in only the flour that the liquid will accommodate. Otherwise, you have to stir all the flour when there is only a little liquid in the bowl.

Adding the flour to the liquid is the easier procedure. Even carefully working your way around the bowl as described above, you're going to have to bash up the flour a bit, and that will develop gluten, which will make the batter rubbery. You might be worried that you'll create a lumpy liquid if you try to merge a little flour with a lot of milk. Don't be concerned about a few lumps: you won't notice them once the batter's cooked.

It's not worth getting out a piece of electrical equipment to make a batter. An electric whisk will give the flour a more aggressive going over than is good for it.

3 • Resting. You don't get much chance to relax when you cook a roast meal, but both the meat (after it's cooked) and the batter (before) should. Resting a batter gives the process that will take place during cooking – the swelling of the starch molecules – a head start, and as a result you'll produce a more cohesive batter. Stir the batter a little when you're ready to use it, to disperse any lumps that have settled on the bottom: it may, because the starch has swelled, have thickened slightly.

4 • Hot fat. To check if it is hot enough, drop in a teaspoon of batter. It should sizzle fiercely, because you want the pudding quickly to form a crusty base. Don't use a thick oven dish: it won't produce a crispy batter.

If, at the end of all this, your beef has been out of the oven for 40 minutes, stay calm. It will be perfect.

STEWS

We used to call them casseroles. They were 1970s dinner party staples; you could cook them from all-colour recipe paperbacks that Marks & Spencer sold for less than a fiver. But the term 'casserole' has come to seem, like prog rock and the bubble perm, rather naff.

They're stews now. 'Stew' has a pleasingly unpreten-tious sound; in the same spirit, we prefer to invite people for 'supper' rather than dinner. Not everyone, though, accepts stew as the generic description of meat (or fish) cooked in a bath of liquid. The meat in a stew, Shaun Hill (*How To Cook Better*) asserts, is not browned before simmering; if it were, it would be part of a braise. This is not a distinction that many others make, even though the most famous dish with stew in the title – Irish stew – does not include browned meat. Richard Olney was not aware of a no-browning rule when he wrote *Simple French Food*: 'Nearly all stews,' he advised, 'belong to the branch commonly (some say incorrectly) known as sautés' – the meat and vegetables are given an initial colouring in fat.

The difference between a stew and a braise, according to several writers other than Shaun Hill, is that a stew contains more liquid. If you don't brown the meat or vegetables, but cover them instead with water or marinade and cook them in a covered pot, you're preparing a daube. If you follow that procedure, but then make a sauce with a roux and the cooking liquid, and thicken that sauce further with cream and egg yolks, you're making a blanquette. Dishes such as coq au vin and boeuf Bourguignon, as well as carbonnades and navarins, are stews of the sauté type.

Sorry. This theorizing won't get the dinner – supper, I should say – ready. I'm going on about the terms because they describe three basic techniques, on which every stew recipe is an improvisation.

HOW TO MAKE IT

Lamb or beef, for 6
1.2kg cubed meat[1] – SEE WHY YOU DO IT
Oil for frying
4 onions, chopped
2 celery sticks, chopped
2 garlic cloves, chopped
Glass white wine
Stock (see pp51–9)
Herbs (parsley, bay leaf, thyme)
Salt
150g mushrooms, sliced
Large knob of butter

Heat a heavy frying pan or ridged grill pan over a medium to high flame. In a bowl, toss the meat with just enough oil (about 1 tbsp, probably) to coat it – you can do the tossing with a spoon, but may find it easier with your hands. When the pan is really hot, put in a batch of meat without crowding the pan. The undersides of the pieces should brown in less than a minute. Turn them, to brown another side, and transfer to another bowl. Repeat the process, until all the meat is browned. You'll find that each batch browns more rapidly than the previous one.[2]

Warm about 2 tbsp more oil in a heavy casserole, and add the onions, celery and garlic, stirring them over a medium to low heat until they soften and take on some colour; it may take 20 to 30 minutes.[3]

Pour in the wine, and allow it to simmer for a couple of minutes.[4] Tip in the browned meat, with any juices the pieces have disgorged, and pour in cold stock barely to cover the contents of the casserole,[5] along with herbs and a little salt. Put the lid on.

Put the stew into the centre of a gas mark 1/140°C oven. After an hour, check it; if nothing is happening, leave it, but check it again every half an hour. You want the gentlest possible simmer.[6] If the liquid is bubbling too fast, lower the heat and/or put the casserole in a lower part of the oven. Don't cook the stew beyond the point at which the meat feels tender when you prod it with a knife or fork. After the pot has come to a simmer, that point may arrive in an hour (lamb) or hour and a half (beef). If you have time, leave the stew to rest for half an hour after taking it out of the oven.[7]

Strain the sauce through a sieve into a saucepan. Separate the meat from the vegetables and herbs, return it to the casserole, and put the lid back on; push down gently on the vegetables and herbs to expel the juice into the saucepan, and then discard them.[8] Spoon off the fat that rises to the surface of the sauce, or dab it off with paper towels.

The sauce will probably be more plentiful, and thinner, than you want. Turn on the heat under the saucepan, bring the sauce to a boil, and simmer it vigorously to reduce it. As it reduces, it will become slightly viscous, owing to the starch that the vegetables have imparted. You may need to start stirring it, so that it doesn't stick, and to turn down the heat, so that the sides of the pan do not burn; you also need to taste it, to ensure that the flavour does not become too concentrated. In particular, you have to watch out for saltiness. It's best to salt the stew very moderately at first, adding more salt later if you need it.

Meanwhile, sauté the mushrooms in some butter (see p162).

Once the sauce has arrived at a volume and consistency

that please you, return it, with the mushrooms, to the casserole, and reheat, covered, very gently. When it is thoroughly warmed through, serve.

All that effort just to make a stew? As I said in the Introduction (p5 onwards), there are plenty of more abbreviated stew recipes, and they'll turn out fine. But cookery books, which are only too happy to give you complicated sets of instructions for fancy dishes, rarely pay the same attention to detail when it comes to guidance on simple ones. You can follow a recipe to the letter, and find that you've got too much sauce, or too little; that the contents of your casserole take the best part of an hour to reach simmering point in the oven, throwing out the timings; that the stew either bubbles too furiously, or not at all; that the meat has not tenderized, or has become overcooked. It's not your fault. Ingredients, utensils and cookers behave differently. Cooking involves improvising to cope with such contingencies.

I apologize for the vagueness of my advice on sauce reduction. But I have to leave this process to your judgement. If I'm serving a stew with rice, or boiled potatoes, I might reduce the sauce for only a short time, and dish out the meal in bowls. Sometimes, I prefer a richer stew, with meat in a coating and small puddle of concentrated sauce.

I have come to prefer to leave out flour from stews. It thickens the sauce at the expense of flavour, and can produce an unattractive congealing effect on the plate. (I rarely make gravy – see p205, above – with flour either.) If you'd like to include it, you can choose from two methods. First, after you've tossed the meat with oil, toss it with as much flour as will give it a fine coating. Browning it with the meat adds flavour, though at the expense of some of its thickening qualities. Or, second, you could stir the flour into the softened vegetables before adding the wine and then the stock, stirring as you go as you would when making a béchamel. Remember, a tbsp of flour will thicken half a pint (284ml) of liquid.

Some cooks thicken stews at the end of cooking with cornflour or arrowroot. I think that these agents add a null quality, but in that view I differ with a more prestigious judge, Raymond Blanc.

You could leave out the wine, or use a different alcohol: brandy or calvados, bubbled until they lose their harsh, spirity edge (see p236); beer or cider.

Add some lardons or chopped-up bits of streaky bacon. Fry them, very gently, first; the vegetables will have some bacon fat to fry in.

To the sauce, add tomatoes: either fresh ones, skinned (see p194) and chopped (almost all recipes tell you to deseed them, but I never bother), or tinned. In carbonnade of beef, the meat is cooked either entirely in beer, or in a mixture of beer and stock; stout works well. Tomato ketchup is a good complement to a rich, fatty cut of beef such as oxtail.

Try Worcester or soy sauce; remember that both are salty. Lamb goes well with lemon and orange – include peelings of zest; and it goes well, too, with spices such as cumin and coriander, which you fry with the vegetables.

Ingredients added towards the end of cooking, as the mushrooms are here, are known as the **garnish**. Other nice garnishes include baby onions (softened gently in butter), lardons of bacon or pancetta (also fried), and croutons – the easiest way to make these is to toss cubes of bread in olive oil, spread them out on a baking tray, and put them in the oven until golden.

A less la-di-da kind of garnish is the **dumpling**. Usually dumplings contain just over two parts self-raising flour to one part suet: Fergus Henderson's recipe (*Nose to Tail Eating*), which makes plenty of dumplings for 6, gives 100g suet, 225g self-raising flour, salt and a beaten egg: you mix them together, adding a little water to bring the mixture to the consistency of a sticky dough, shape them into walnut-sized balls, and cook for about 10 minutes. You could poach or steam them, in a covered pan containing water; or you could ladle into the pan some of the liquid from the stew, put in the dumplings, cover, and simmer.

That kind of dumpling seems to be the right match for a beef stew. Dried beans or chickpeas suit lamb particularly well. Cook them apart (see p144 and 167), and add them at the same time as you would any other garnish.

Some fresh herbs could go into the stew just before serving: parsley, say, or tarragon, or oregano.

If you don't want a garnish, but have skimmed the sauce, consider adding a knob of butter, away from the heat, just before serving. It will enrich the dish.

Boeuf Bourguignon belongs to the family of stews in which the meat has an initial bath in a marinade. There are many strict and elaborate recipes, so let us take comfort – as we have to do so often – in Elizabeth David's observation that 'Such dishes do not, of course, have a rigid formula.' You might put your beef into a marinade consisting of a bottle of red wine, a chopped onion, a tablespoon of olive oil, some herbs, and salt. (The wine might be a Burgundy or a Côtes du Rhone. I'm sure you don't want to use an expensive one; but don't use one you would find undrinkable.) Leave it overnight. The next day, remove the meat and pat it dry. Follow the basic recipe as above, using the marinade, sieved, as the cooking liquid, and topping it up with stock if necessary. Try to include bacon in the initial preparation; baby onions are the best garnish.

WHY YOU DO IT

1 • What meat? The meats for stewing are the 'tough' cuts: among them chuck or stewing steak, oxtail, shin of beef; or shanks, shoulder, middle neck and scrag end of lamb. They come from the bits of the animal that get the most exercise, and that have muscles with a strong reinforcment of connective tissue. This tissue, consisting largely of a protein called collagen, is very difficult to chew; but, subjected to slow cooking, the collagen breaks down into succulent gelatine.

The proteins in muscle fibres toughen up, expelling all their moisture, if cooked for too long. That's why we fry or grill lean cuts such as fillet steak as quickly as possible. If we stewed them for any length of time they would become – in spite of their liquid surroundings – unbearably dry. In tough cuts, the fibres dry out too, of course; but they become looser, and their gelatinous lubrication gives them a tender feel in the mouth.

2 • Browning meat and the 'sealing' theory. You are searing the meat, but most definitely not, in spite of the surprisingly resilient popularity of the term, 'sealing' it. The notion of frying meat in order to 'seal in' the juices is to cooking what bloodletting is to medicine. That sizzling noise in the pan? It's the water from the meat being vaporized as it reacts with the pan and the hot oil. Have a look at a piece of browned meat. Does its surface appear to you to be water-impermeable?

The high heat required for browning actually causes rapid fluid loss. Cooks usually think that the sacrifice is worthwhile, though, in exchange for the flavours from the Maillard reactions, the complex and tasty results of the collision at high temperatures of carbohydrates and amino acids in proteins.

To reduce the risk of burning the oil, add oil to the meat, not to the pan.

Make sure that the pan is thoroughly hot, and do not throw in too much meat at once. If you do, you will lower the temperature, and the water that should have vaporized instantly will flood the pan, boiling the meat rather than browning it.

You could brown the meat in the casserole dish, before you soften the vegetables – or after, provided you have removed every trace of the vegetables from the dish first. My method is easier, I think.

3 • Softening the vegetables. The onions and garlic, particularly, lose their harshness during this process, becoming sweeter. You may fear that the garlic will burn while the onion is softening; but it doesn't seem to. If you're worried, add it a couple of minutes before this stage is complete.

I have been a bit vague about the amount of oil you will need. You want just enough to ensure a layer of oil between the vegetable and the bottom of the casserole, to stop them catching. Start with a little, and add more as necessary.

Many recipes include carrots at this stage. As I remarked in the Stocks chapter, I think that the flavour of carrots in a sauce can lose its freshness after long cooking. If you want the sweetness of carrots in a stew, add them half an hour before the end of cooking.

4 • Reducing the wine. To lose its harsher notes. Of course, the sauce will be subject to a good deal of evaporation anyway. But you may not want the wine to take up a significant percentage of it.

5 • Drowning the meat. So long as it is submerged in liquid, the meat will not have to endure a temperature much higher than 100°C (although some of the contents of the sauce may raise the boiling point).

6 • Gently does it. The heat and rollicking motion of the liquid in a rapidly boiling stew will drive out moisture, as well as fat and gelatine, from the meat; you'll be left with lumps of dry fibre. Always remove meat when you're boiling a sauce to reduce it.

With the surface of the stew showing only the mildest evidence of disturbance, the temperature of the liquid may be several degrees below the 100°C boiling point (if you put it on the hob and turned up the heat, it would take a little while to come to a proper boil). The loss of those few degrees, as well as of a good many agitated bubbles, will give you a much more beguiling result.

I used to bring stews to simmering point on the hob, then put them in the oven. Harold McGee has reformed me. You just have to learn not to be dismayed to find, particularly if you have a heavy casserole, that the stew you put in a 140°C oven an hour ago is showing little evidence of progression.

You needn't worry. There is progress. Meat starts to cook when it reaches about 50°C; its collagen starts to dissolve at

70°C. Even at below 50°C, the gentle heat is weakening the connective tissue, reducing the time the meat will have to spend at temperatures that will dry it out.

However, you may not have the luxury of an extended deadline. If you need the stew to be ready in three hours or under, start it off at gas mark 6/200°C, but turn down the dial when the liquid simmers.

As soon as the meat in a stew is tender, stop cooking. It won't get any more tender, but it will get more dry.

7 • Resting. The meat will reabsorb some liquid as it relaxes in the cooling stew. An overnight rest will do your stew even more good than a half-hour one: the flavours will develop; and fat will rise to the surface and solidify, so that you can remove it easily.

(I should point out that if you leave overnight a stew cooked with stock, then reheat it, you are going against the Food Standard Agency's advice that you should not reheat a stock twice – see p55. But I have never poisoned myself, or heard of anyone being poisoned, in this way, and I'm happy to carry on reheating my stews.)

8 • Disposable vegetables. Keep them if you like. But they have already done their job, imparting flavour to the sauce.

Two much simpler stews

Tough cuts of meat such as middle neck of lamb and oxtail may not need to be entirely submerged in liquid, because they can remain tender in spite of exposure to the hot air inside a casserole. In which case, you do not need to brown them first – they brown as they cook. You turn them as they do so.

LAMB STEW

For 4

900g middle neck
3 onions, roughly chopped
1 head garlic, separated into unpeeled cloves
1 orange, cut into quarters
2 bay leaves
Sprig rosemary
1/2 chicken stock cube (optional)
Salt
1 tbsp oil

OXTAIL

For 4

900g oxtail
3 onions, roughly chopped
1 garlic clove, chopped
1 tbsp tomato ketchup
1 tsp soy sauce
1 tsp nam pla (fish sauce)
1 star anise
1/2 stock cube (optional – I use a chicken cube here too)
Salt
1 tbsp oil

In each case, toss the ingredients in a heavy casserole, and put them, covered, in a gas mark 1/140°C oven for 3 to 4 hours, or until the meat is tender. You can lower the heat once the contents of the casserole are simmering. Turn the meat occasionally, as it browns.

You're starting these ingredients from cold, and you may find, if you have a heavy casserole, that they take far too long to heat up at this low setting. Particularly if time is short, set the dial at first to gas mark 6/200°C, but turn it down as soon as the stew begins simmering.

I find that the stock cube makes a difference, giving an extra depth of flavour to the sauce.

You have added no extra liquid, but you will end up with quite a lot of sauce. You could strain it, skim it, and reduce it (perhaps, in the lamb stew, squeezing into the sauce the softened garlic from the hulls), as in the more fiddly stew above. If you really do want this meal to be as straightforward as possible and to serve the stew as it comes, you should probably take into account the fattiness of the sauce by serving it with plain rice or boiled potatoes.

CHICKEN STEW

You brown the pieces of chicken first, but more gently than you do beef or lamb: chicken is more delicate; also, it doesn't throw out so much water, so you don't need a high heat to ensure that it fries rather than stews.

A chicken stew is not the same as a chicken sauté (see p266).

Here, off the top of my head, is a recipe: it has something in common with chicken Basquaise, plus a goulashy note.

HOW TO MAKE IT

For 4

4 chicken legs[1] – SEE WHY YOU DO IT

Olive oil

Salt

2 onions, chopped

2 garlic cloves, chopped

1/2 glass white wine

400g can tomatoes, or 6 fresh tomatoes, skinned and chopped (see p194)

2 red peppers, quartered, deseeded and sliced

1 tsp paprika

2 tbsp sour cream

Add just enough oil – maybe no more than a dstsp – to a casserole to provide just a thin layer, and put the dish over a low to medium heat. Salt the chicken legs, and fry them, two at a time if necessary, for about 5 minutes each side, or until lightly browned.[2] Remove them to a plate.

The chicken will have exuded some of its own fat. You could take the view that chicken fat is not particularly delicious, and throw it away (but not down the sink, where it will congeal). Or, thriftily, you could keep it and use it to soften the onion and garlic for 10 to 15 minutes, until mellow. If you've thrown away the chicken fat, soften the onion and garlic in olive oil.

Pour the wine into the dish, scrape and stir into it any

sediments from the browning process you can find, and reduce this liquid by about half. Add the tomatoes (you can bash up tinned ones with a wooden spoon as they heat in the pan), peppers (which, not having a harsh flavour, don't need pre-softening), paprika and a little more salt. Return the chicken to the casserole too, bring everything to a gentle simmer, cover, and continue to cook for about 45 minutes on a very low heat; or put the casserole into an oven at gas mark 1/140°C, or at a setting that will maintain a gentle simmer.

At the end of cooking, remove the chicken to a warm plate, pour the sauce into a saucepan, then put the chicken back into the casserole and cover. (You could return it to the oven, turned off.) Put the saucepan on to a high heat to reduce and thicken the sauce, which will need stirring if it is not to catch. When the sauce has the consistency you want, turn off the heat, check the seasoning, and stir in the cream; pour the sauce over the chicken. Rice is the obvious accompaniment; I'd also enjoy this stew with mashed potato.

VARIATIONS

In this recipe, the liquid might not cover the chicken – so perhaps you should call it a braise. As a moistening agent, you could use, instead of tomatoes, some chicken stock, in which case you might want a thickener for the sauce. Dust the chicken with flour before frying it. Or you could braise the chicken in just a little liquid – that provided by the wine, for instance, or even by a tablespoon of vinegar; the chicken will exude juices too. At the end of cooking, finish the sauce with a little butter or cream.

Add dried porcini mushrooms and their soaking liquid to the braise. Add some sautéed fresh mushrooms at the end. Add fresh herbs: thyme and tarragon are particularly good.

Coq au vin was ubiquitous in the days when we referred to stews as casseroles, and, perhaps for that reason, it has fallen out of fashion somewhat. A revival is due. The bird should be a cockerel, but you'll probably have to make do with chicken.

With one, jointed chicken, use a whole bottle of Burgundy or Côtes du Rhone, as well as a double measure of brandy. Follow the basic beef or lamb stew recipe above: flour, salt and brown the chicken, and remove to a plate; fry onion, celery and garlic; throw in the brandy, and reduce it until it almost disappears; pour in the wine; return the chicken, minus the breast portions, to the casserole. Bring the casserole to a simmer and cook for 45 minutes to an hour, or longer if you like your chicken really tender; add the chicken breasts and cook for a further 15 minutes. Strain the sauce, skim and reduce it as in the basic stew recipe on p239: you want quite a concentrated reduction.

For a garnish, fry lardons, as well as mushrooms and/or baby onions. Coq au vin is often served with fried bread: one slice, crusts removed, for each person. As you probably know, bread in a frying pan absorbs a heartstopping quantity of butter; you could lightly butter one side, put the other on a buttered piece of foil, and bake it. It's not the same, though.

Nigella Lawson has a recipe for **chicken cooked slowly with lemon and garlic**. You cut the lemons (two) into pieces, mix them with a jointed chicken, a head of garlic cloves, seasoning, 150ml white wine and some olive oil, and bake everything slowly in a roasting pan covered with foil. Half an hour before the end of cooking, you uncover the pan and turn up the oven temperature, to brown the meat. The drawback of this method, I think, is that browning meat at the end of cooking can expel the last drops of moisture from it; an initial searing expels some juices but leaves behind plenty, some of which, with luck and care, you can retain. So I'd brown the chicken first, in a frying pan; then mix it with the lemon, garlic and salt in a roasting pan, cover the pan tightly with foil (or simply use a heavy casserole) and bake at gas mark 1/140°C for about an hour and a half. You might want to strain the sauce into a saucepan and reduce it.

The best of all chicken and garlic dishes, and one quite unlike any other stew or sauté recipe, is **chicken with 40 cloves of garlic**. My favourite way of cooking it is the simplest. Put a chicken, cut up into pieces (or just buy thighs and drumsticks), in a heavy casserole. Separate, but do not peel, the cloves of 4 heads of garlic (it has to be good garlic; and, of course, it has to be a good chicken), and add them, along with salt, and thyme if you like. Pour in a generous glug of olive oil – a good 150ml. Mix everything well with your hands.

It's important that the casserole should have a good seal, so that the juices don't evaporate. Make a paste with flour and a little water, forming it into a thin sausage and placing it around the rim of the casserole; jam the lid on top. Put the casserole in a gas mark 4/180°C oven for an hour and a half; you could turn down the heat to gas mark 1/140°C at the point when you guess that the chicken is cooking – but don't break the seal to have a look. As you pull off the lid at the table, you release a fabulous aroma. You can eat the garlic, squeezed out of the hulls, with the chicken, or spread it on bits of toast or on potatoes.

WHY YOU DO IT

1 • Forgiving legs. The legs and wings of chicken take best to stewing, and are tolerant of most treatments. You don't have to handle them with the kid gloves you need when dealing with beef and lamb (see p229): slow-heating and complete submergence of the meat are not essential.

If I buy a whole chicken for a stew or a braise (or for grilling), I get my butcher to joint it. The task takes him about a minute; it takes me at least a quarter of an hour of effortful hacking.

For reasons explained above (see Roast chicken, p203, and What meat?, p234), the lean breast meat cooks quickly and then dries out even more quickly; give the breasts just 15 minutes or so in the pot. Because you cannot interrupt the cooking of chicken with 40 cloves of garlic, you might prefer to use thighs and drumsticks only for that dish.

2 • Browning reactions. What you're browning is mostly skin; but that will contribute to the flavour. The skin protects the flesh, which on contact with high heat dries out and toughens unappetizingly.

Pork stew

Most cuts of pork are too lean to stew or braise successfully. The exceptions include belly and spare rib chops – not to be confused with spare ribs (see p219).

A BELLY PORK BRAISE

For 4

900g boneless belly pork, cubed
6 onions, roughly chopped
Olive oil
As many garlic cloves as you like
2 lemons
Salt

Sear the meat as in the lamb and beef stew recipe above (p229). Soften the onion in the oil (1 to 2 tbsp) in a casserole dish. Add the garlic and the lemon, cut into quarters, and tip in the pork. Season.

Or: simply toss all the ingredients together, as in the simple stews, above (p238). Any pork exposed to the air in the casserole will gradually brown. Turn it as it does so.

Cook, covered, in a gas mark 1/140°C oven for 2 to 3 hours, or until the meat is meltingly tender. You can turn down the dial once the contents of the casserole are simmering. No degreasing and skimming here: the fattiness, cut through with lemon juice, is an important component of the dish. Serve it with something plain: rice and a green salad, perhaps.

A SPARE RIB
PORK BRAISE

HOW TO MAKE IT

For 2
2 spare rib chops
1 tbsp olive oil
1/2 glass white wine
Chicken stock
Salt
Small pot (142ml) double cream
1 tsp mustard

Sear the chops as in the lamb and beef stew recipe above (p229). Transfer to a casserole; throw in the wine and a ladleful of stock. Add a little salt. Put, covered, in a gas mark 1/140°C oven, or simmer very gently on the hob, for 60 to 90 minutes, or until the meat is tender. (If the pork is not entirely submerged, turn it halfway through cooking.) Remove the meat from the casserole, and pour the liquid into a saucepan; return the meat to the casserole, putting the lid back on. Reduce the liquid over a high heat until it is a little syrupy, then add cream and mustard, and bubble until it thickens. Check the seasoning. Pour back over the meat, and serve.

CASSOULET

Most cassoulet recipes are highly elaborate. Richard Olney in *The French Menu Cookbook* gives 30 ingredients and 4 stages of cooking. We home cooks, in love with cassoulet but too lazy, or busy, to prepare the restaurant-standard version, call again in our defence what Elizabeth David said about boeuf Bourguignon (p233): in essence, that individual cooks interpret provincial dishes in their own ways. You're not sullying the spirit of the dish if you simplify it – quite the opposite.

More complicated and, I suppose, authentic recipes include lamb or mutton as well as pork. Three kinds of meat (and meat fat) are one too many for me.

HOW TO MAKE IT

For 6
500g dried haricot beans
6 slices boneless pork belly, rind removed
2 onions, 1 of them studded with 2 cloves, the other chopped
4 garlic cloves, 1 of them chopped
6 Toulouse, Cumberland or other coarse sausages
6 portions of confit of goose, or confit of duck[1] – SEE WHY YOU DO IT
1 tbsp or more of the fat from the goose or the duck
3 tomatoes, skinned and chopped (see p194)
Chicken stock
Salt, herbs
Breadcrumbs

Soak the beans in cold water overnight, drain them, cover them in fresh water to a height about 4cm above the top of the beans, bring to the boil, skim off the scum that rises, add the clove-studded onion with the 3 unpeeled cloves of garlic and the rind from the pork belly, and simmer, partly covered. It's hard to predict how long the beans will take to soften; start checking after an hour, and top up the water if the topmost beans are in danger of being exposed. When they are nearly cooked, uncover the pan and allow the liquid to reduce and thicken – it will be a little sludgy. There is going to be another cooking stage, so the beans needn't be completely soft. (For further advice on dried beans, see p144.)

Tinned beans, as I've said earlier (p73), are fine, but somewhat duller-tasting than dried ones, and with a more mealy texture. If you use them here, you'll miss out on the flavoured stock that cooking dried beans would have given you, but you'll produce a perfectly nice meal in far less time. Mix them with the pork, tomato and onion stew (see below), before covering with stock.

Meanwhile, heat the goose or duck fat in a heavy casserole, and brown the belly pork as in the basic lamb or beef stew recipe on p229; a medium heat should do. Set the meat aside on a plate, turn down the heat, and add the sausages, keeping them moving in the casserole until they are browned in various places. Set them aside too. Add the chopped onion until it is softened (use more fat if necessary), then add the clove of chopped garlic. Let the garlic soften, then add the chopped tomatoes.

Drain the beans, reserving the liquid; keep also the garlic and the pork rind, but throw away the onion. Chop the rind into small squares, and add it, with the beans and the garlic (which will probably melt out at some point, leaving people to find the bare husks on their plates), to the onion and tomato mixture in the casserole. Add the belly pork slices, burying them among the beans. Pour in the bean liquid, along with enough stock to come to the level of the topmost beans. Add salt to taste, and herbs if you want them (bay and thyme, perhaps). Cover the surface with breadcrumbs.

Put the casserole, uncovered, in a gas mark 2/150°C oven. Check it after an hour: the mixture should be bubbling gently, and have a crusty surface. If it's bubbling too fast, or not at all, adjust the temperature accordingly. If all's going well, push the breadcrumbs into the bean mixture, and add another layer of them to the surface. Check again after another half an hour. Is there another crust? And is the liquid reducing and thickening? If yes to both, slice the sausage, and add it and the pieces of duck or goose to the beans;[2] you don't have to submerge them completely. Cover with another layer of breadcrumbs, and cook until this too has formed a crust. Serve.

That was quite hard work. Still, the only accompaniment a cassoulet needs is a green salad.

WHY YOU DO IT

1 • The confit. Advice on making your own is beyond the scope of this book. You can buy goose or duck confit (the latter is easier to find) at smart shops. Do keep all the fat, which makes particularly splendid roast and fried potatoes.

2 • Add the sausages and confit late. Sausages go flabby and dull if stewed for too long. The duck or goose is already cooked; what it would add to the cassoulet if there from the beginning would not compensate for what it would itself lose.

LAMB DAUBE

This is a simplified version of the Avignon daube that appears in Richard Olney's *Simple French Food*.

HOW TO MAKE IT

For 6
1.5kg shoulder of lamb, cubed

For the marinade
1 bottle dry white wine
1 onion, chopped
1 carrot, chopped
3 garlic cloves, crushed
1 tbsp olive oil
Salt
Parsley, bay leaf
Zest from half an orange

To add to the daube
6 shallots, chopped
3 garlic cloves, chopped
4 tomatoes, skinned and chopped
Salt

Put the marinade ingredients, with the exception of the wine, with the lamb in a non-reactive dish. Pour over enough wine to cover. Put on the lid, and leave overnight.

The next day, lift out the lamb and put it in a heavy casserole; pour the marinade over it through a sieve, discarding the solid ingredients. Add the chopped shallots, garlic, tomatoes and salt (remember that the marinade contained salt). Mix everything together, and pack it in quite tightly. If the marinade doesn't cover the ingredients, do not worry at this stage, because the meat and vegetables will generate a good deal of further liquid. Bring the contents of the casserole, uncovered, very slowly to a simmer on top of the hob; you can allow an hour or more to get to this point.[1 – SEE WHY YOU DO IT] Skim off the scum as it rises. Cover, and put in a gas mark 1/140°C oven, or at a setting that will maintain a very gentle simmer, until the lamb is tender – start checking after another hour. Skim off what fat you can from the surface; or, better, leave overnight (in the fridge), and lift off the solidified fat the next day.[2] If you have kept the stew overnight, warm it up gently, but do not cook it again.

WHY YOU DO IT

1 • Slow-heating. See Gently does it (p236). Raw meat produces a good deal more scummy material as it heats in liquid than does meat that has been browned. Warming the daube on the hob, and leaving the casserole uncovered, allows you to skim off the scum as it rises. In a covered, simmering pot, the scum would merge with the liquid. It won't do you any harm, though.

2 • Runny liquid. Reducing the sauce in a daube would be as beside the point as doing it when you make a Lancashire hotpot or Irish stew. You serve the meat and vegetables moistened with their cooking liquid. Noodles or rice are ideal accompaniments.

BLANQUETTE

Again, adapted from *Simple French Food*.

HOW TO MAKE IT

For 4

4 chicken legs or 1kg stewing veal or 4 spare rib pork chops

Water

1 onion stuck with 2 cloves

Herbs

Salt

Butter

Flour

Small pot (142ml) double cream

2 egg yolks

Put the meat in a casserole, add water barely to cover, and bring slowly to a simmer, uncovered, on the hob, skimming as you do so (see Slow-heating, p251). Add the onion, herbs of choice and salt, and simmer, very gently, on the hob or in a gas mark 1/140°C oven (see Gently does it, p236). When the meat is tender (45 minutes to an hour for the chicken, perhaps 90 minutes for the veal), take it off the heat, leave it to rest for 10 to 15 minutes (see Resting, p210), and remove it to a warm plate.

Sieve the liquid, discarding the onion and the herbs, into a measuring jug, and work out how much flour and butter you'll need to make a velouté – one that remains quite thin, with only a little body (see Velouté, p48). Return the meat to the casserole, and cover. Make a roux, and follow the béchamel procedure. Check the seasoning, and pour the sauce back over the meat. Mix the egg yolks and the cream, and season with pepper if you like. Ladle a little of the hot sauce into this mixture – you're getting the eggs used to the high temperature they're about to meet. Pour the eggs and cream into the stew, stirring carefully and warming gently. Don't let the sauce boil: you'll end up with a liquid containing bits of scrambled egg. When the sauce thickens some more, serve. (Olney's recipe includes sorrel, which you stew in butter, then incorporate into the cream and egg mixture.)

Curry

What I'm getting at in this book is that most home cooks are not interested in authenticity; they want to acquire templates of cookery techniques and recipes, so that they can be confident of putting together, from ingredients to hand, something that will taste nice. I reiterate that point here in a probably doomed attempt to mitigate the offensiveness of taking a liberal position on the creation of a spicy stew. That heading, 'curry', is ill-advised for a start: its use as a generic term for spicy food is a British habit, not an Indian one.

Buy whole spices when possible. Warm them gently in a dry saucepan, until they give off a toasted aroma. Grind them with a pestle and mortar, in a herb mill or in a coffee grinder. They need a little cooking, in fat, to lose their powdery taste; but you have to do it with care, because they burn easily. Don't fry finely ground spices such as chilli powder and turmeric: they will certainly burn. Be aware, too, that the aroma from these toasted spices as they stew with the meat will invade your home, and linger there.

I am going to leave the composition of the spice mixture to you. For 4 people, you could use a 2 tbsp perm of several of the following: asafoetida,^{SEE NOTE BELOW} cumin, coriander, chilli powder, dried chilli, fresh chilli, black peppercorns, fenugreek, cloves, turmeric, cardamom (use just the seeds, or put whole pods into the stew), ginger (fresh, grated), mustard seeds, mace, nutmeg. Some recipes include cinnamon. If you can find curry leaves, you could add, say, half a dozen to the frying mixture. Stir in coriander leaves, plus more fresh chilli, at the end of cooking.

Start by frying onions – I'd include 3 or 4 – in enough oil, in a casserole dish, to keep them well lubricated; or, if you have it, use ghee (Indian clarified butter). Get the onions to brown (see p164), then add 2 chopped cloves of garlic, and let that take some colour too. Add your spice mixture, and cook it gently for about a minute.

Meanwhile, brown your meat, as you would in a beef, lamb or chicken stew (pp229 and 239). Tip it into the spiced onions, and add water or stock. Don't use wheat flour to thicken the liquid – that really would give the curry an inauthentic quality. If you want a thickener, and can find chickpea flour (besan flour), try that. At the end of cooking, reduce the liquid if you want – you might need to concentrate the flavour. But don't skim off the fat. Count the calories elsewhere; this is too flavourful to waste.

For a Thai-style curry, use 2 tbsp of a paste consisting of chopped lemon grass, green chillies, shallots, lime leaves (if you can find them; or use some lime zest), coriander leaves and stems, garlic, and ginger or galangal (the latter, again not easy to find, is preferable).

This is a fresh-tasting mixture that benefits from shorter cooking. One possibility: fry the paste for a minute, add cubed chicken breast, and pour over a tin of coconut milk, or a carton of coconut cream. Add a little salt. Simmer, uncovered, stirring occasionally; the chicken may be cooked through in 10 minutes. If the liquid is too thin, boil it hard (with the chicken removed) to reduce it, and return the chicken to warm through again. Add coriander leaves at the end.

Note: use asafoetida powder sparingly. Even 1/6 tsp will exert an influence, both earthy and zesty, on your dish. Its pungent odour will linger, too. I love it.

Mince

Among the most unrealistic of all recipe book instructions are the ones that tell you to fry some onions, then add mince to the pan and brown it. At this medium temperature, water gushes out of beef, pork or lamb mince, but does not vaporize; it floods the pan, stewing rather than frying the meat. If you turn up the heat to evaporate the liquid and hasten the browning reactions, you'll find that, once the mince and onions start to fry, they catch and burn.

There are three techniques that will work. The first is to evaporate the liquid at high heat, and then to turn the dial right down, allowing the mince to brown gradually. The second is to keep the heat low throughout, waiting patiently for evaporation and the browning reactions to take place – the process takes 45 to 50 minutes, in my experience. The third is to form the mince into patties, oil them lightly, and sear them as you would the cubes of meat in a stew (p229). You are browning only the portion of meat on the surfaces of the patties, of course. When you incorporate them in the dish, you break them up with a wooden spoon.

Even the best pre-prepared mince will be nowhere near as good as mince the butcher prepares for you. Ask for coarsely ground chuck or stewing steak, or shoulder of lamb, or belly pork. Don't stew lean mince; it's no more suitable for long cooking than is fillet or sirloin steak. Use it for hamburgers.

The meat in a ragu alla Bolognese (Bolognese sauce) is usually beef, or a mixture of beef and pork, and is not browned, according to several authorities, Anna del Conte among them. The meat in a shepherd's pie is lamb; make it with beef, or beef and pork, and you've got a cottage pie.

I use fresh or tinned tomatoes and/or tomato paste in a ragu, but prefer the sweetness of ketchup in a shepherd's pie, offsetting it with the rich saltiness of Worcester or soy sauce.

RAGU ALLA BOLOGNESE

HOW TO MAKE IT

For 4
Olive oil

100g pancetta or streaky bacon, preferably unsmoked, finely chopped

1 onion, chopped

1 celery stick, chopped

2 garlic cloves, chopped

350g beef mince, or mix of beef and pork mince

100ml white or red wine
1/2 400g tin of tomatoes, or 4 fresh ones, skinned and chopped
1 dstsp tomato paste
Chicken stock
Salt

Put a little oil (about 1 dstsp) in a heavy saucepan or casserole, and fry the bacon gently; it will release its own fat. After about 5 minutes, add the onion, celery and garlic; you probably won't need any more oil. Allow them to soften and go golden for 10 minutes or so, stirring regularly. Turn the heat right down, and add the mince, breaking it up with a wooden spoon. When it has separated (as I say above, you don't need to brown it), pour in the wine; give the alcohol time to simmer and reduce a little before adding the tomatoes and tomato paste, with enough stock to come up to the surface of the meat. Season, and bring the sauce very slowly to a simmer, uncovered, on the hob. Continue to simmer, at the lowest heat possible (with a heat disperser under the pot if necessary) and stirring from time to time, for about 2 hours, until the sauce has reduced and the meat is tender.

You could include 100ml double cream or milk with the liquid you add to the ragu. Many recipes include carrot with the onion, celery and garlic. Carrot undoubtedly adds sweetness to the sauce; but, as I say elsewhere, I don't particularly like finding flavourless bits of it in my meal. If you have no stock, don't use a cube, which would impart a flavour quite unlike anything you want to find in a pasta sauce; try a bought carton or jar of stock, or (used cautiously) a tin of beef consommé.

Use this sauce with spaghetti, tagliatelle or any pasta you like; use it (without cream or milk, which will be in the béchamel) in lasagne (p125). A simpler version, with lamb instead of beef (though of course you can use beef if you want), forms the basis of a moussaka.

MOUSSAKA

HOW TO MAKE IT

For 4

Olive oil

3 onions, chopped

3 garlic cloves, finely chopped

500g lean minced lamb (or beef)

150ml white or red wine

1 tbsp tomato paste

4 aubergines, cut in 2 horizontally, then cut vertically
   in 1/2cm slices

Béchamel sauce (p46), made with 500ml milk,
      57g butter, 57g flour

20g Parmesan cheese

2 eggs, beaten

Put a layer of olive oil in a casserole, and soften the onion and garlic in it over a gentle heat, until the onion is golden – 15 to 20 minutes.

Meanwhile, form the mince into 6 hamburger-shaped patties, and lightly coat them with oil. Put a ridged grill pan or heavy frying pan on a medium to high heat; when the pan is very hot, brown the patties, perhaps in 2 batches of

3. They should take no more than a minute a side. Transfer to a plate.

Pour the wine into the casserole with the softened vegetables, and simmer for a couple of minutes. Add the tomato paste and a little salt; a pinch or two of cinnamon, if you like it, works well in a moussaka. Tip in the browned patties of mince. Heat the contents of the casserole gently, breaking up the patties so that the uncooked meat in the centre gets heated; simmer until there is no liquid left. Test the seasoning.

The last time I made a moussaka, I didn't fry the lamb: I simply added it to the softened onions and garlic, broke it up, poured in wine, added tomato paste with salt and a pinch of cinnamon, and simmered this stew until the liquid had evaporated. It may have been just as good.

Brush the aubergine slices with olive oil, season with salt, and bake in a gas mark 6/200°C oven for 20 minutes to half an hour, or until tender. I'm sorry: I expect you'll have to do this in 2 batches.

Make quite a thick béchamel (see p46). Stir in the Parmesan. I think that you should add some nutmeg too. Allow the sauce to cool before stirring in the eggs – so that the eggs don't curdle.

Put a layer of aubergines on to the bottom of an ovenproof dish. Pour all the meat over the top, and cover with the rest of the aubergines. Cover the whole lot with the enriched béchamel.

Bake at gas mark 4/180°C for about 30 minutes, or until the topping has puffed up, thanks to the eggs, and turned golden.

You can eat moussaka hot, warm, or at room temperature.

SHEPHERD'S OR COTTAGE PIE

For 4

Olive oil

100g pancetta or streaky bacon, preferably unsmoked,
    finely chopped

1 onion, chopped

1 celery stick, chopped

450g lamb mince, or beef mince, or beef and pork mince

100ml white or red wine

1 tbsp tomato ketchup

1 tsp Worcester or soy sauce

Chicken stock

Salt

900g maincrop potatoes (King Edward, Maris Piper
    or Desirée, for example)

50g butter

A combination of the ragu and moussaka recipe.

As with the ragu: put a little oil (about 1 dstsp) in a heavy
saucepan or casserole, and fry the bacon gently; it will release
its own fat. After about 5 minutes, add the onion and celery;
you probably won't need any more oil. Allow them to soften
and go golden for 10 minutes or so, stirring regularly.

As with the moussaka: form the mince into 6 hamburger-
shaped patties, and lightly coat them with oil. Put a ridged
grill pan or heavy frying pan on a medium to high heat; when
the pan is very hot, brown the patties, perhaps in 2 batches of

3. They should take no more than a minute a side. Transfer to a plate.

Pour the wine into the casserole with the softened vegetables, and simmer for a couple of minutes. Tip in the mince. Add the ketchup, Worcester or soy sauce, and enough stock (about 250ml, possibly – remember that it will reduce and thicken) to make a stew with a thick and clinging consistency. It shouldn't resemble soup. Err on the side of caution: you can add more stock later, but to reduce the sauce you'll have to put the whole stew (you can't very well drain the sauce from the mince) on a very gentle heat, uncovered, and wait for the desired reduction to take place. Be careful, too, with the salt, because Worcester and soy sauce are very salty. Break up the patties with a wooden spoon.

Put the casserole, covered, in a gas mark 1/140°C oven (see Gently does it, p236). As with any stew, you're aiming to warm it up slowly towards the gentlest possible simmer; check on progress from time to time, and adjust the temperature accordingly. The time in the oven should be 2 to 2 1/2 hours.

Peel the potatoes, cut them into slices about 1.5cm thick, cover with cold, salted water, bring to a gentle simmer, and cook until tender. Drain them, and return them to the hot saucepan to dry. Mash them; a smooth purée doesn't seem necessary for a shepherd's pie, so use a hand-held masher rather than a food mill or a potato ricer (see Mashed potato, p177).

Spread the potatoes on top of the mince, in the casserole or, if that seems too big, in some other oven dish, and dot the butter on top. Bake in a gas mark 1/140°C oven for about 20 minutes, then brown the surface under the grill.

Garlic isn't really a shepherd's pie kind of vegetable, but do include it if you like. Mushrooms work well, though: add them when the onions and celery have been cooking for 10 minutes or so, and stir them around until their juices have poured out and evaporated. Delia Smith's recipe includes 75g of swede, chopped up small. She also suggests half a teaspoon of cinnamon (not to my taste), and a tablespoon each of parsley and thyme.

Most recipes for Bolognese sauce and shepherd's pie tell you to use meat stock, but that's not something I have around the house very often. For shepherd's pie, you could use a stock cube. Half a cube, I find, adds savouriness, whereas a whole one is too assertively artificial. Or try stock in a carton or bottle, or a tin of consommé. If your stock is good, and if you don't fancy ketchup or Worcester sauce or herbs, you could make a perfectly delicious shepherd's pie with just the mince, the vegetables and the stock.

A cheese crust on the potato is a nice enhancement. Add a topping of grated Cheddar, or pecorino, or Parmesan, before putting the pie in the oven.

MEATBALLS

HOW TO MAKE THEM

For 4
200g beef mince
200g pork mince

2 heaped tbsp grated Parmesan
3 tbsp breadcrumbs
Pinch of cinnamon
1 garlic clove, salted and mashed
Zest of 1 lemon
1 egg, beaten
Salt and pepper
Olive, sunflower or groundnut oil for frying

In a bowl, mix the ingredients, apart from the oil, with your hands. I find that the breadcrumbs, and to a certain extent the Parmesan, help to retain a loose texture – without them, the meatballs become compacted when cooked. The egg is not always essential, but can help to bind the ingredients.

Pick up small portions of the mixture, and roll them gently into balls between your hands. I like golfball-sized portions.

Warm a heavy frying pan over a medium heat, pour in a little oil, and fry the meatballs in batches, turning once. The heat should be sufficient to brown each side in about a minute. Remove the browned meatballs to a plate.

Make a tomato sauce (p196–7). Drop in the meatballs, and simmer gently, covered, for about 15 minutes. If the tomato sauce is already thick, you may find two problems: that you cannot submerge all the meatballs in it (which is why you cover the pan); and that it sticks to the bottom of the pan as it cooks further. You may need to finish the dish in the oven, instead.

Serve hot, warm, or at room temperature.

You could also make a **meat loaf** with the above ingredients. Chop an onion, fry it gently in a tbsp or two of olive oil until golden, and tip it with its oil into a loaf tin (mine is 7cm x 16.5cm). Gently pack the meat mixture on top, and bake at gas mark

4/180°C for an hour. You may find some juice floating on the surface. Throw it away. Run a knife round the edges of the loaf, and turn it out on to a plate. Again: serve hot, warm, or at room temperature.

A BOILED DINNER

HOW TO MAKE IT

For 6
900g piece of gammon
3 onions, peeled
1 leek, tough green parts removed, washed (p160)

Some gammon is very salty, and may need soaking in several changes of water for about 24 hours. Ask your butcher. I doubt that gammon from the supermarket will need this treatment.

Put the gammon in a stockpot. Cover with cold water. Bring slowly to a simmer; following the same principle that cautions you to heat a stew slowly (see p236), you can allow a good half an hour for the bubbling to start. Skim off the scum, and throw in the vegetables. Simmer, uncovered and on the lowest possible heat (with just a few bubbles breaking the surface), for a further hour.

Remove the gammon from the stock, carve and serve with mustard, mashed potato (p177), and cabbage. You can leave the meat to rest for a while before carving; or allow it to rest in the stock after the heat has been turned off.

The purpose of the onions and leek is not to be eaten, nor to flavour the gammon (they won't), but to add flavour to the stock. The liquid may look insipid, but it will form the base of delicious soups.

You can also use the stock to make a sauce. Pour a couple of ladlefuls into a saucepan, boil it to reduce it to about half a dozen tablespoons, check that it won't be unbearably salty, and pour in a 284ml pot of double cream, with 1 tsp mustard. Allow this sauce to bubble and thicken a little, and serve it with the gammon. You might want boiled potatoes, or rice, instead of the mash.

Or make a béchamel-type sauce with half stock, half milk (see p46–7). Say, 150ml stock, 150ml milk, 28g butter, 28g flour – or larger quantities of all these ingredients if you like. Throw in a chopped tablespoon of parsley, or stir in some cooked spinach (p193).

My butcher sells bacon hocks, each with enough meat for about 3 people. They're cheap. The meat can withstand long simmering: I usually give it a couple of hours.

Sautés

'When dealing with chickens,' Richard Olney writes in *Simple French Food*, 'a sauté is not a stew, because a true sauté, it is claimed, never contacts a liquid during the cooking process.' (Actually, the book has 'a true stew', but that must be a misprint.) Some authors ignore this distinction, or perhaps disagree with it. Anne Willan, in *Classic French Cooking* (an old Sainsbury's book), offers a sauté of chicken with tomato: you brown the jointed chicken, then make a sauce with shallots, garlic and tomatoes; then you put the chicken in the sauce to finish cooking. You might as well call that a stew. On p239, I did. It's related to such dishes as chicken Basquaise, with tomatoes and peppers, or chicken cacciatore, with tomatoes and olives. What I'm describing here is a dish in which you brown the chicken, remove it from the pan, pour away most of the fat, and make a sauce that accompanies the meat.

CHICKEN SAUTÉ

HOW TO MAKE IT

For 4
4 chicken legs
Olive oil, or butter/oil mix
200g button mushrooms, sliced
1/2 garlic clove, chopped
Butter for sautéing the mushrooms
Small glass white wine

Put a small knob of butter and a little oil in a heavy frying pan (large enough to contain the 4 pieces of chicken) over a low heat. Season the chicken pieces with salt and (if you like) pepper. Cook gently for 25 minutes, uncovered.[1] – SEE WHY YOU DO IT Turn over; the skin should be browned and crisp. Cook for a further 10 minutes.[2] (Rearrange the chicken in the pan if some pieces are cooking faster than others.) While the chicken cooks, sauté the mushrooms and garlic with some butter in a separate pan (see p162).

Remove the chicken to a warm plate. Pour away the fat from the chicken pan (not down the sink, where it will solidify), and deglaze it (p55) with the wine, stirring and scraping to get the tasty residues in the pan into the sauce. Bubble the wine until it reduces by half (p267), add the mushrooms, and check the seasoning. Serve the chicken with this mushroom sauce.

VARIATIONS

Liquids for sauces: red wine, brandy, cider, calvados, sherry, marsala, vermouth …; or vinegar (2 tbsp, reduced by half); or chicken stock. You could add cream (a 284ml pot) to these deglazing liquids, or deglaze the pan with cream alone; let it bubble a little to thicken.[3] If using alcohol or stock alone, enrich it once it is reduced with a small knob of butter, added away from the heat.[4]

Other garnishes: shallots (onions would be too assertive for this dish – although small, sweet ones would be good), also softened in butter with some garlic; dried porcini, soaked in tepid water for half an hour or so and warmed through with the deglazing liquid (and their own flavourful soaking liquid); sliced, sautéed courgettes; roasted asparagus tips.

Herbs: tarragon is particularly good with chicken. Or thyme, or basil. Add them to the sauce when it is ready.

ACCOMPANIMENTS

You shouldn't want fried or roast potatoes with a creamy sauce: that gives you two kinds of mouth-coating fat to eat and digest. It's an indication of the crudity of my taste that I'm sometimes happy to face this combination.

Boiled new potatoes are ideal; if you've got plenty of sauce, boiled maincrop potatoes work well too. Mash (p177) or gratin dauphinois (p188) is nice if the sauce does not include cream.

If you've eaten a starter including starch, you could leave out the potatoes and go simply for a green vegetable, or perhaps a ratatouille (p173).

WHY YOU DO IT

1 • Open pan. Conscientiously sticking to the Olney 'no cooking the sauce with the chicken' rule, how are you going to make sure that the chicken is cooked through, rather than just browned on the exterior? Olney suggests transferring it from the frying pan to a covered casserole, and finishing it in that. Richard Ehrlich recommends a sauté pan with straight sides and a lid. You brown the chicken, then lower the heat and cover the pan.

My method above comes from *Keep It Simple* by Alastair Little and Richard Whittington. It also has a satisfyingly pure quality: the chicken is simply sautéed (unless you insist that the term, derived from the French verb meaning to jump, involves moving something around a frying pan), not stewed or steamed. On the downside, the juices it exudes evaporate, or get poured away with the fat. If you decide to cover the pan, fry the chicken first in only a dessertspoon of oil, so that the juices that accumulate in the covered pan are not too oily. You do not want to pour them away. Reduce them, if necessary, in a saucepan. If you're using alcohol or vinegar, remove the chicken briefly to a plate, pour the alcohol or vinegar into the now-dry sauté pan, turn up the heat, and bubble to reduce by about half. Tip this liquid into the sauce in the saucepan, return the chicken to the sauté pan, cover it to keep warm, and finish the sauce in the saucepan with cream, stock or garnishes, as above. (You wouldn't be able to reduce the wine/vinegar so efficiently if you added it to the saucepan liquid.)

You could cook the chicken in the oven, and make a sauce in the roasting tin. But that is roast chicken with gravy.

2 • Is it cooked? If you're not sure, put a skewer into the thickest part of the thigh. Juices will run out. They should be clear. If there's any trace of red in them, carry on cooking.

3 • Curdling cream. In my experience, good double cream does not curdle in these circumstances. Delia Smith writes somewhere that crème fraîche never curdles; that is not my experience. Factory-made crème fraîche curdles when I cook it, without fail. But rich, thick, and much more delicious farm-produced crème fraîche never does.

Cream goes from runny to stiff quite quickly, so keep an eye on it. Having some extra stock, to thin the sauce if necessary, is useful. Don't add any more alcohol: if it is not subjected to a proper reduction, it will impart a horribly raw, acidic flavour.

4 • Splitting butter. Butter added to boiling liquid will immediately separate into its constituent parts of fat and water. Let the sauce cool a little, and you have a chance of allowing the butter and liquid to blend into a creamy, rich emulsion. Don't heat it again.

Other sautés

Fried steak with a sauce made in the same frying pan (see p273) is, of course, a sauté of this kind. You may have had it prepared at your local restaurant by a waiter with a small burner, a bottle of brandy and a smarmy manner. To do it at home, you need a gas cooker, a frying pan that you can manipulate easily, and some of that waiter's confidence. Fry the steak, pour in the brandy, turn up the gas, and tip the pan so that the flame hits the brandy and ignites it. Watch out. When the flame dies down, pour in cream, and remove the steaks immediately, even if you want to bubble the sauce a little longer to thin it – the meat will toughen quickly if it stews.

However, that's if you want to show off. The sauce will work just as well – and may be nicer – if you bubble the brandy in a separate pan to reduce it before adding it to the steak (see p267).

You can also treat veal steaks and pork tenderloins in this way.

Frying and grilling

Meat fried or grilled is fast food. You cook it at a high heat to get the flavourful Maillard or browning reactions (see p234); and you cook it briefly, in order to avoid expelling all its juices. A pork chop is hard to get right. It is thick, and may need a good 15 minutes in the frying pan or under the grill; but after that length of time, it is invariably dry. (See p276 for an attempt at the solution to this problem.)

Meat for frying and grilling comes from the animal's most relaxed bits, which have done the least exercise and which therefore don't have toughened connective tissues. From a cow, those bits include the sirloin and the fillet; from a sheep, the loin and best end of neck; from a pig, the loin again; from poultry, the breast (although every bit of a chicken can be grilled or fried). Browned and cooked to the point at which it is just done (in the cases of steak and lamb, most gourmets prefer some rareness), this meat is firm but juicy; tender too. Cooked beyond that point, it suddenly expels all that juiciness, and becomes shrivelled and leathery.

A ridged grill pan, or a barbecue, is best for grilling meat. Overhead grills do not sear the meat as satisfactorily; they also seem to dry it out more.

Salt

Don't salt meat before you cook it, you're advised: you'll draw out the juices. Meat consists of about 75 per cent water. A dusting of salt on raw meat will not have a significant effect on that percentage; and, as the meat cooks, it undergoes a drastic water-reduction process anyway. That sizzling as the meat hits the pan or the grill is the sound of water, instantly vaporizing; the high temperature, not the salt, is responsible for it. When the meat rests after cooking it exudes more juices, whether it is salted or not.

Salt added to meat before cooking will not penetrate much, but will add savour to the browned crust. It's an effect worth having, in my opinion.

Marinades

So, yes, you can add salt to a marinade (see above).

Books on barbecuing and grilling can lead you to believe that, in order to be interesting, grilled meat must first be bathed in some exotic compound, for at least 6 hours. I find the Hugh Fearnley-Whittingstall argument persuasive. In essence, it's this: the acid in marinades will mess with your meat; if you've got a decent piece of meat, why would you want that to happen? If you want a lemony, or vinegary, taste, marinate with lemon or vinegar just before you start grilling.

Without acid, however, marinades scarcely penetrate the surface of the meat. Punctilious cooks use syringes to inject flavoursome compounds, but that kind of technique is not for you and me. We're happy if our marinades give a coating of flavour, enhancing the finished dish; and therefore we need apply them only when we're ready to cook.

Here are various types of marinade you can rub over your meat. Use some oil with all of them: it provides lubrication, flavour and a certain amount of protection, both for the dry ingredients of a marinade and for the meat – oil gets very hot, but not as hot as a naked flame.

Herbs: thyme, rosemary and oregano are good; also, for pork, sage.

Spices: see Curry on p253. Mix these with oil before coating the meat (you could do the same with the herb marinade). They are less likely to burn with an oil coating. An alternative is to mix the spices with yoghurt. A yoghurt marinade, if given time, will work on the meat, which in this case you want to be imbued with as much spiciness as possible. Harissa (p134), thinned with oil (olive, sunflower or vegetable), is an alternative fiery coating. Or a jerk marinade: a mixture of Scotch bonnet chilli (it's very hot, so remove the pith unless you're brave) with, say, 4 spring onions, a tsp-sized piece of ginger, 1 tbsp honey, 1 tsp cinnamon, salt to taste, and the juice of a lime. Whizz them up in a food mill, and stir in enough olive and sunflower oil to form a paste. Watch out when you're grilling: the honey burns easily.

Garlic: crush it with some salt and rub this paste over the meat.

Barbecue marinade: it might consist of tomato ketchup, soy sauce, mustard, vinegar, crushed garlic, salt, pepper and oil. Honey or brown sugar is optional, with the proviso stated above: if your grill is fierce, you may end up with blackened food.

You can also use this marinade to coat slow-cooked meat, such as spare rib chops or belly pork (or chicken wings from a stock – see p51). One method is to poach the meat, very gently, for an hour, or until tender; drain, coat in marinade, and put on the barbecue or in a gas mark 6/200°C oven until the marinade becomes pleasingly sticky – 5 to 10 minutes (or perhaps 15 to 20 if the meat is heating up again from cold). What you hope is that the marinade will protect the meat enough to stop it drying out. Or you could bake your chops or belly pork, smothered in marinade, very gently in a roasting tin covered with foil, before grilling it. The disadvantage here is that the meat will shed a lot of liquid, giving you a tin full of diluted sauce, which will be an unsatisfactory marinade for the final, grilling stage.

You can poach spare ribs, for about an hour on a very gentle heat in an uncovered pan. Drain them, chop them into individual ribs, marinate them, and bake or grill them as above. It's a good idea to line your roasting dish or gill pan with foil.

Frying and grilling steak, or hamburgers

Warm the frying pan (I use my heavy cast-iron one), over a medium to high flame. Coat the steak in a little sunflower, vegetable or groundnut oil (which withstand higher temperatures than does olive oil), and salt it. The oil assists in the transmission of heat, and at the same time protects the steak from the very high heat of the pan. If you applied the oil to the pan rather than to the meat, you might burn it.

Don't crowd the pan with steaks: you'll lower the temperature, halting the browning reactions (see p234). After one minute, turn over the steak, which should have browned. (If it is brown in much less than that time, you're probably cooking it too fast; if it's not brown, the heat is too low.) After another 60 seconds, turn it over again.

Now, how thick is your steak? Hugh Fearnley-Whittingstall says that a 3cm-thick steak will be rare in 3 to 4 minutes, and medium rare in 5 to 6.

The problem is that your steak is already browned after 2 minutes; leave it for another minute on each side, and the surface meat will burn and dry out. Turn down the dial, and you may find water entering the pan: the meat, which the browning has not sealed (see p234), continues to expel its moisture, which the pan is not hot enough to vaporize. When the meat stews, it dries out very rapidly: water has this paradoxical effect, because it is such an efficient cooking medium.

One answer is, once the initial browning has taken place, to turn the meat very regularly – every 20 seconds or so, until it is done to your liking. It gets enough heat to continue cooking, but not enough to cause it to burn. Or put it in a gas mark 6/200°C oven for the remainder of the cooking time. To judge whether it's done, you might have to dig into it and have a look.

Put the steak on a warm plate to rest for a few minutes (see p210). It will exude some moisture, but will nevertheless be juicier and more tender.

You could make a simple sauce, as you would with a chicken sauté (see p266).

To grill, rub the steak with a little oil and seasonings before putting it on a barbecue or ridged grill pan. Follow the frying procedure, above.

You fry or grill a hamburger in the same way. Some people like to select pieces of steak and get their butchers to mince them, but I usually make do with the excellent steak mince my butcher offers. I season it with salt and pepper, and, heretically in the views of some, add beaten egg to it (1 egg for 450g mince), to help it cohere. I form it into patties of about 150g each.

Grilling lamb

I prefer to fry steak, but to grill lamb, perhaps because lamb is often fattier; also, the grilling flavours complement the meat particularly well. Thicker cutlets or lamb steaks may require 5 to 10 minutes of cooking to get to medium rare; but you can turn down the heat under your pan without risk of the stewing that steaks suffer under the same treatment. On the barbecue, start the cutlets in a place of high heat, turn over after a minute, brown the other side, then move them to a place where the embers are less fierce.

Frying and grilling chicken

You can fry or grill any parts of a chicken. But the legs take about 25 minutes to cook; they are quite difficult to manage on a barbecue, and almost impossible to grill successfully on a ridged pan: the skin and outer flesh burn before the heat penetrates the interior. You could use an overhead grill, which cooks chicken much more satisfactorily than it does other meats; or you could bone the legs, creating a thinner cut. Make an incision starting from the thin end of the leg, and follow the line of the bone to the thigh end; prise away the flesh, and chop off the tendons. The absence of bone will not cause the chicken to be less tender, unless you overcook it. (It's worth boning thighs, but not drumsticks.)

A thick piece of chicken breast on the grill or barbecue may develop an overcooked outside and a rare interior. Finish it in the oven; or (my preferred option), cut it into thinner pieces before cooking; or cube and skewer it, then grill.

The easiest way to cook chicken wings, unless you're happy fiddling about with 20 pieces of meat or more on your grill, is in the oven. I like to get them very tender. Try mixing the wings with a barbecue sauce (see p272) and baking them for 30 to 40 minutes at gas mark 5/190°C, turning once. They respond well to slow cooking too: 75 to 90 minutes at gas mark 2/150°C.

You can pan-fry chicken legs (or just thighs) slowly (see Chicken sauté, p266). Or make a seasoned crust: mix flour with black pepper, cayenne pepper and salt, and toss the chicken pieces in it thoroughly, perhaps in a paper bag if you have one large enough. Transfer to a pan containing hot (but not smoking – i.e. burning) oil. You need a medium heat, to crisp the crust; so perhaps it's a good idea to bone the chicken first. (Dipping the chicken in egg or buttermilk before flouring it can give you a claggy, rather than crisp coating.)

Grilling pork

A pork chop is the cut of meat for non-cooks. All you do is shove it under the grill, turn it once, and it's ready in 15 minutes. Serve with plenty of mustard or bottled sauce, to compensate for the dryness of the meat.

The Food Standards Agency advises that pork, along with chicken, be cooked through. We shouldn't, according to this official advice, leave in it the pink centre that in beef or lamb represents tenderness and juiciness. But a pork chop is particularly lean: it lacks the fatty lubrication that will give an impression of tenderness even when the muscle fibres are dry. An overcooked chop is an uningratiating meal.

The best way I have found to retain some tenderness in a chop is to fry or grill it for a minute on each side (see steaks, above), and then give it a further 15 to 20 minutes, depending on thickness, in the oven (gas mark 6/200°C). Hugh

Fearnley-Whittingstall has a nice recipe (for 4): fry a whole head of separated, unpeeled garlic cloves in a layer of olive oil in a large pan for a few minutes; add the chops and brown for a minute on each side; remove the chops and garlic, and place in a preheated oven dish, with the thinner ends of the chops pointing upwards to expose the fat; throw away the oil; deglaze the pan (see p55) with a glass of wine or cider, reduce the liquid by half, and pour it over the chops; season, and put in the oven. The drawback can be that garlic does not always soften in this time. If in doubt, poach it for 10 minutes first.

Sausages

They may be an unassuming foodstuff, but sausages are difficult to cook well. The challenge is to get them nicely browned without causing them to burst and lose their juices.

HOW TO COOK THEM

Melt a small knob of butter in a heavy frying pan. Put the pan on a heat disperser on one of the back rings of your hob, and place the sausages in it. The heat should be so low that, after 25 minutes, it has just begun to brown the undersides of the sausages. Turn them; the other side will brown more quickly – perhaps in 15 minutes. Turn again. And, after another few minutes, again. And so on (see Frying and grilling steak, p273). At this heat, it may take an hour to cook through the sausages properly.[1] – SEE WHY YOU DO IT If the browning is progressing faster than the cooking, you'll need to turn the sausages frequently from the moment the browning process starts.[2]

You might also grill or bake them.[3]

WHY YOU DO IT

1 • An hour to cook a sausage? Even for a sausage sandwich? If you have the time, it's worth taking it. They don't split; they keep their juices; and they remain tender.

2 • Defusing bangers. Sausages came to be called bangers because of their high water content; as they heated up in the pan, the water would boil and cause them to explode. The preventive measure was to prick the skins, to give outlets to the liquid. We have less need to defuse the better sausages that are more widely available nowadays, and we don't like to encourage the loss of juiciness. But sausages still split quite easily. A gentle heat preserves the succulence.

3 • Grilling and baking. Among grilling tools, a barbecue works better than an overhead grill. A barbecue's heat may appear to be very fierce, but it is usually less aggressive than a grill flame, which has a tendency to dry out sausages and toughen their skins. Another advantage of the barbecue is that, as fat leaks from the sausages, it spits back at them, forming a tasty crust. Don't put sausages on a ridged grill pan: they'll stick. The oven has its advocates. I have sometimes produced succulent sausages from the oven; more often, I have got dry ones, and I haven't worked out how to guarantee the best results. The problem, I think, is that the oven's heat is too strong; but, if you turn it down, you don't get much browning. My best advice is to warm a little olive oil or lard in a roasting pan, add the sausages, and put them in a gas mark 5/190°C oven for 15 minutes, then turn them over, and cook them for another 10. If they're already nicely browned when you first look at them, turn down the oven a couple of settings. Don't add the sausages to scorching hot oil: you'll be lucky if they don't split.

Olive oil or lard is my preferred fat here: butter will burn in the oven, and I am not so keen on the taste of sausages cooked in sunflower or vegetable oil.

I'll keep trying the oven, because it's often easier to cook sausages that way than to fry them; but gentle frying, in order to retain as many juices as possible, is my preferred method. No, you don't have to spend an hour doing it. But, if you want to cook sausages at a higher heat, turn them regularly, or the skins might char and split before their contents are properly cooked.

Bacon

Cheap bacon is pumped full of water. As you fry or grill it, a slimy brine floods the pan; the meat contracts and curls. A good deal of the salt remains behind in the bacon, too, in concentrated form. You get a piece of meat that is dry, rubbery and salty. You may think that the quality of bacon matters less than that of other meats, because all you want it for is a cheap and cheerful sandwich with lots of ketchup and mustard; but good bacon – which is still pretty cheap – is worth the extra money.

I prefer streaky to back bacon. Back is more expensive, because it is leaner; but because it is leaner, it is inclined to become very dry when you cook it.

I prefer to fry bacon, in a little oil and over a low to medium heat, or to put it on to a ridged grill pan, than to cook it under an overhead grill, which produces a drier and tougher rasher.

A fishmonger – if you're lucky enough to have one within reach – can be the most daunting of food shops. The conceptual gap between the creature on the slab and a plate of food seems, to the inexperienced cook, too large to bridge. Those scales, fins and gills don't advertise edibility; nor do those shells and pincers.

Fish, with the possible exception of the Japanese ones that kill you if you don't prepare them properly, need not be alarming. They will respond to a brief subjection to one of the basic cooking methods: poaching, steaming, baking, grilling, frying. (Oily fish such as mackerel are not suited to poaching or steaming – that's about the only exception you need to know.) The fancy part of the process is making a sauce, but the home cook may want to leave elaborate sauces to fancy kitchens. It seems pretentious to make something complicated when, cooked simply, it is unimprovable.

You could take home one of those daunting fish, without even knowing what species it was, and simply wash it, season it, brush it with a little oil, and put it in a gas mark 6/200°C oven for 15 minutes or so; it is likely to be delicious. You could put your unnamed fish under the grill. Or you could wrap it in foil, with some chopped shallots, parsley and a few tablespoons of white wine. Or get the fishmonger to fillet it for you. Keep the heads and the bones, and make a stock with them (p57). Poach the fillets in the stock; then reduce the stock and swirl in some butter and cream to make a sauce. Or fry the fillets, accompanying them with the fish stock sauce.

The biggest headache is the size of the fish. One person will eat an entire sole; how, then, are you going to fry, grill or even bake enough for six? You might manage it for four, using two frying pans, frying a sole in each, and keeping them warm in a low oven while the other two cook. But you'll wish you were serving only two, as you will when cooking a good many fish dishes. If I have a larger number to feed, I look for fleshy fish that will fit in my oven or under my grill, or I make a stew (p300), or I poach a salmon.

Do bear in mind that many of the fish mentioned in this chapter are under threat. As I write, the Marine Conservation Society's list of species one can eat in good conscience includes coley, mackerel, mussels, oysters, pollack, red mullet, sardines or pilchards, scallops, and lemon sole. Other species may be thriving only in certain regions. There are at present, for example, good stocks of Atlantic cod – a species that has been subjected to notorious overfishing – in the north-east Arctic, but not in the North Sea. Haddock stocks in the North Sea are healthy again, and are also good in the north-east Arctic, but are in poor shape off the west coast of Scotland.

POACHED SALMON

You need a fish kettle – a long, narrow pan with a lid. Unless you have a long, central ring on your hob, you can put the kettle over two rings. Or, if you have a roasting pan deep enough to contain a salmon and water to cover it, try that.

The traditional fish-poaching liquid is court bouillon – a kind of vegetable stock made with water, aromatic vegetables, vinegar or wine and salt. I have cooked a salmon in court bouillon and in plain, salted water, and cannot say that the latter was less interesting. I suspect that the influence of flavourings in water on a large fish over the course of half an hour's cooking is minimal. Salt, though, does penetrate the flesh, and tenderizes it.

Put the salmon in your kettle, cover it with water, add a generous portion of salt, and turn your ring or rings to low/medium. When bubbles start to appear, turn down the heat. 1 – SEE WHY YOU DO IT From this point, a 1.5kg salmon will take 20 minutes to half an hour to cook. Test with a knife at the thickest point. You can look at the flesh, which should have turned opaque; or you may be able to tell by the feel of the fish, which should be flaky rather than firm.

Serve, hot or cold (cold is more common), with boiled new potatoes and mayonnaise (p41).

VARIATIONS

Nigel Slater recommends this method: you bring the fish slowly to a simmer in water containing 50g of salt for each litre (that's a lot), put on the lid, turn off the heat, and leave overnight to eat for lunch the next day.

For 2 people you could poach salmon fillets or steaks. Cook them briefly, in hot water: bring enough salted water to cover them to the boil, turn down the heat, and slip in the salmon. Test after 3 minutes: they'll probably be ready.

1 • Gently does it. Salmon, being fatty, is a forgiving sort of a fish. Nevertheless, it and other fish require the care you give to meat when cooking, and for the same reason: they dry up and toughen if heated for too long. A poaching liquid will cook a fish most sympathetically if kept at a temperature some way below boiling point.

POACHED WHITE FISH

By white fish, I mean fillets of such fish as cod, haddock, hake, plaice, sole, turbot, brill and sea bass. You can use any kind of pan, including a frying pan, large enough to hold the fish; and, because they cook so easily, you don't have to cover them completely in liquid. Simply turn them halfway through to immerse the exposed sides. Flat fish fillets can be rolled or folded to fit the pan.

For 2

Bring a couple of ladlefuls of fish stock (p57) to a simmer in a saucepan or frying pan. Turn down the heat to the gentlest simmer. Slide in the fish and cook until the flesh becomes milky, turning halfway through if the slices are not completely immersed. Even a chunky piece of cod or haddock won't take more than 5 minutes in total. Lift out the fish using a slotted spoon, transfer to a warm plate, and put it in the lowest part of a warm oven.

Turn up the heat under the stock, and boil until you have just a few tablespoons of syrupy liquid. Throw in a small pot (142ml) of double cream, and bubble until thickened. Add 1 tbsp chopped parsley and, away from the heat, a walnut-sized knob of butter. Check the seasoning, and serve the fish with the sauce.

VARIATIONS

You could poach the fish in milk, and use that as the basis for the sauce. Many parsley sauces are béchamels (see p46), so you could use the poaching liquid to make one of them (if you use stock, the sauce is called a velouté); but I find a flour-based sauce a bit heavy as an accompaniment to delicate-fleshed fish, except in fish pie (see p288).

Leave out the cream; just enrich the reduced stock with a little butter, added away from the heat to prevent splitting. Sauté some mushrooms, with a little garlic if you like (see p162) and add them to the reduced stock. Sharpen the sauce with a squirt or two of lemon juice.

STEAMED FISH

You can steam a fish or a fish fillet – the same kinds of fish that you would poach (see above) – as you would vegetables, in a steaming basket (see p24) placed above salted boiling water in a saucepan. Or you could put a rack inside a roasting pan, and steam your fish, covered tightly with foil, either on the hob or in the oven. A whole fish may take 12 to 15 minutes; a fillet may be ready in less than 3 minutes. The flesh turns white, and acquires a flaky, rather than taut, texture.

There, you have a steamed fish. Even I, an advocate of simplicity, can see that it needs a little embellishment. An eggy sauce, such as mayonnaise (see p41) or aioli (p42) or hollandaise (p44), might suit. Or try an oriental theme.

For 2

1 garlic clove

2 spring onions, coarse green parts removed

1 tsp ginger

1 dstsp sunflower, groundnut or vegetable oil

1 dstsp vinegar

1 dstsp soy sauce

1 dstsp sesame oil

Here's the schedule for a very quick meal: put a steaming basket in a saucepan, pour in water beneath it, add salt, put a lid on the pan, and put the pan on the heat.

Cut up the garlic, spring onions and ginger, and get ready

the other sauce ingredients. By this time, the water should be boiling; place a couple of fish fillets in the basket, and cover.

Warm a frying pan or saucepan over a medium to high heat.[1 - SEE WHY YOU DO IT] Add the oil, and when it is hot (but before it smokes), add the garlic, spring onions and ginger, and stir them around quickly until the onion wilts a little – no more than 30 seconds.

Add the vinegar, and let it evaporate for a minute.[2] Remove from the heat, and stir in the soy sauce and sesame oil.[3]

By this time, the fish may be ready; remove the fillets to warm plates, and pour the sauce over each.

WHY YOU DO IT

1 • Warming the pan first. Or else you might overheat the oil. See p34.

2 • Evaporate the vinegar. Reducing vinegar – and alcohol – removes some of its harsh notes. See p34.

3 • Delicate sesame. You have to be very careful if you fry with sesame oil – it burns at a low temperature. Here, you're using it as a flavouring rather than as a frying aid.

You can also half-poach, half-steam fish in milk in the oven. It's a useful method for preparing the fish for a fish pie: the milk doesn't boil too rapidly, as it might do in a covered pan on the hob, and then provides the basis for the sauce.

FISH PIE

For 4

650g smoked haddock

560ml milk

10 black peppercorns

1 bay leaf

900g maincrop potatoes (King Edward, Maris Piper
   or Desirée, for example)

57g butter

57g plain flour

Handful of parsley, chopped

50g more butter

Put the haddock fillets in an ovenproof dish, and set the oven at gas mark 4/180°C. Heat the milk gently in a saucepan with the peppercorns and bay leaf; when it starts to bubble, pour it over the haddock, cover the dish with foil, and put it in the oven for 5 to 10 minutes. The fish has got more cooking to come, so it doesn't need to be cooked through; all you need now is to get it to a stage at which you can flake it.[1] – SEE WHY YOU DO IT

Or: put the haddock into a saucepan, cover with the milk, peppercorns and bay leaf, bring gently to the boil, turn off the heat, cover the pan, and leave for 5 to 10 minutes.

Peel the potatoes, cut them into slices about 1.5cm thick, cover with cold water, bring to a gentle simmer, and cook until tender. Drain them, and return them to the hot saucepan to dry. Mash them; as when making shepherd's pie (see p260), you may not feel that a perfect purée is essential, so use a hand-held masher if you like. Season. (No milk and no butter at this stage.)

Lift the fish out of the milk, which you should strain and reserve.

Melt the butter in a non-stick saucepan, add the flour to make a roux, cook it gently for a couple of minutes until sandy (but don't let it darken), and gradually add the strained milk to make a béchamel (see p46), stirring continuously. Let the sauce bubble for a minute or so, then throw in the parsley. Check the seasoning; you may not need any salt, because the fish is salty. The sauce should be thick enough to coat the back of a spoon.[2]

Pull the fish off its skin, separating it into decent-sized flakes (it's easiest to do this by hand). Stir the flakes gently into the sauce. Then pour this mixture into a pie dish (pre-warmed in the oven), and cover it with the mash. Cut the butter into little pieces, and dot them on top.

Bake the pie as in the shepherd's/cottage pie recipe: in a gas mark 1/140°C oven for about 20 minutes, before browning the surface under the grill.

VARIATIONS

You don't have to use smoked haddock. Any fish (including salmon, but not including mackerel and sardines) will do; but smoked fish works particularly well.

Add sautéed mushrooms to the sauce (see p162). Or stewed leeks (p160). Or cooked, peeled prawns.

Hard-boiled eggs are nice in a fish pie. The problem is

that they are usually hard-boiled before they go in; then they get 20 minutes in the oven, and overcook. Try this: boil the eggs (one for each person) for 6 minutes; drain them and cover in cold water (to arrest the cooking). Peel them, and bury them gently, whole, in the fish and sauce mixture just before you cover the whole lot with mash. They should finish cooking with the pie. If they're still a little soft, will you mind?

WHY YOU DO IT

1 • Pre-cooking the fish. The fish is about to be immersed in sauce and put in the oven for 20 minutes. Even at this low temperature, it will cook through. But if you simply added skinned, raw, chopped-up fish to your béchamel, it would exude a good deal of its own liquid and thin the sauce to a probably disastrous extent. The pre-cooking allows it to shed some of its liquid, flavouring the milk; it also enables you to skin and flake the fish more easily. You bake the pie at a low oven temperature in order to try to ensure that the fish does not overcook. But the top will not be brown after 20 minutes at gas mark 1, so you need the grill as well.

2 • Thick sauce. The sauce will undergo conflicting processes in the oven. The further cooking will thicken and concentrate it; but any further liquid from the fish will thin it. A sauce that's too thick is preferable to a runny one, so err on the side of thickness.

Pan-fried fish

I sympathize with Elizabeth David in her dislike of the term 'pan-fried'. It seems factitious, created because chefs think that 'fried' sounds a bit downmarket – just as executives think that they appear more dynamic if they 'head up' rather than simply 'head' an organization. 'Waiter, is this fish fried?' 'No sir: it is pan-fried.' (The chef's phrasal verb 'fry off', meaning to fry something to rend its fat or to temper its flavour, is similarly irritating.)

Having got that off my chest, I have to admit that the term creates a useful distinction between fish that is fried in a thin layer of fat, and fish that is immersed in fat completely, usually with the protection of a coating such as batter.

PAN-FRIED SOLE OR PLAICE FILLETS

HOW TO COOK THEM

For 2

Your largest, 28cm frying pan may just be able to contain 4 fillets of fish. Season them on both sides with salt, and pepper if you like, and dredge the flesh sides in a plate of flour, to give them a light dusting.[1 – SEE WHY YOU DO IT] Put the pan on a medium heat, and add enough butter to provide a generous coating to the surface, with 1 dstsp or so of sunflower or olive oil.[2] When the butter is foaming, slip in the fish, skin side down.[3]

The fish won't take long to cook. You might turn a thin

fillet after a minute, and give it no longer than 30 seconds on the flesh side.

Lift the fillets, skin side up,[4] on to warm plates. Pour away the butter and fat. Add more butter – 50g, say – to the pan, and swirl it around until it's melted. Throw in a chopped handful of parsley, and squeeze over the sauce 1 tbsp lemon juice. Pour the sauce on to the plates with the fish.

Spinach (see p193) is an ideal accompaniment; boiled new potatoes, too (p176).

VARIATIONS

Any fish fillets, provided they are not very thick, will be good when given this treatment.

I think that sole, plaice and other delicate white fish need no further embellishment. Salmon, though, will take the addition of strong extra flavourings. Try Nigel Slater's idea: fry the salmon as you would the sole or plaice (see above – the salmon will need longer in the pan), discard the butter, add more, then add a tablespoon or two of rinsed capers, and 1 tbsp red or white wine vinegar. Bubble this sauce for a minute or so to soften the raw, vinegary taste.

You can cook whole flatfish or mackerel in this way, too. There's no point in flouring them first (see p293). Cook them on a moderate heat, to avoid scorching the skin before the inside of the fish is tender. You can fit several mackerel into a large frying pan, but only one flatfish, which one person will happily consume. So you'll need either to cook one sole or plaice at a time, keeping the first one warm in a low oven while you tackle the second, or to use two pans – if your hob is large enough to accommodate them.

WHY YOU DO IT

I • Dredging. Cooks give fish a light dusting of flour in order to protect the tender flesh, and to provide a crispy surface. I'm not sure that the flour does much protecting: Tom Stobart (*The Cook's Encyclopaedia*) insists that it doesn't. So this stage is optional.

2 • Oil with the butter. Elizabeth David tells us that the fish in sole meunière – the proper name for the recipe above – should be fried in clarified butter, which doesn't burn as readily as does the normal stuff. A simpler way to avoid burning is to add oil to the butter. When butter is alone in the pan, the liquid evaporates from it; the solid material in the remaining butter gets very hot, and burns. The addition of oil, which retains its liquid consistency on heating, thins out the butter, and makes burning less likely. The oil will have only a small influence on the flavour, because your sauce will be based entirely on butter.

3 • Skin side down. The skin protects the delicate flesh, and crisps as it fries. Fry the fish on the skin side until only a brief cooking on the flesh side will bring it to readiness.

4 • Skin side up. Is serving fish with the skin confronting the diner a ridiculous restaurant affectation? Not entirely. The skin will go soggy if the fish is turned over. But if you don't want to eat the skin anyway, this serving method is redundant. If you do want to eat the skin, don't pour the sauce over it – unless you like your fish skin soggy.

Fry and bake

Chunky steaks of cod or other fish will spend too long in the pan, and dry out, if you try to cook them through by frying. Instead, you start them on the top of the stove, to crisp the skin, and complete the cooking in the oven.

Melt a layer of butter and oil in a roasting pan, or any other pan that will go in your oven, until the butter starts foaming.

Put in the fish, seasoned with salt and pepper if you like, skin side down. Fry it for a minute or so until the skin starts to crisp, then turn it flesh side down and put the pan in a gas mark 6/200°C oven. The fish may take 8 to 10 minutes to cook; you'll be able to see that all the flesh has turned milky white.

Once you've removed the fish from the pan, you can follow the meunière method above: pour away the cooking fat, and melt more butter in the pan (which will be hot, so you probably won't need to return it to the hob), along with lemon and parsley if you like. A tablespoon of rinsed capers would be nice, too.

The flesh side of the fish will have more protection in the oven if facing downwards than if facing upwards, when it would be exposed to the radiant heat.

You may not want to throw away the butter and oil in which the fish has cooked. As long as they haven't burned, they'll make a fine sauce themselves.

Fish in batter

When I have made battered cod, I've infected the air of the house for several days. The fish has tasted great, though.

Make a batter (according to the recipe on p225) with beer (any kind) as your liquid, but use half the amount that you would in the normal recipe: in other words, 224g flour, 1 egg and 280ml beer. That should be plenty for 2 decent-sized fillets. The liquid is meant to be so thick that only a little of it falls off when you move the fish from batter to oil.

Pudding and pancake batters benefit from a rest before cooking. However, Harold McGee tells us that coating batters for frying will be crisper if used immediately.

In a fryer or deep saucepan, heat enough sunflower or vegetable oil to cover the fish, to 175°C. If you don't have a thermometer, drop in a small piece of bread, which should sizzle vigorously.

Salt your fish fillets; that will make their surfaces sticky. Then dredge them in flour, which will help the batter to adhere. Dip the fillets in the batter, holding them at their thinnest points. Quickly, so that not too much batter falls off, transfer them to the hot oil; but submerge them gradually. They may rise to the surface; if they do, turn them halfway through cooking. An average-sized fillet will take about 10 minutes to cook. The batter should be puffed and crispy; the fish, which has steamed inside its coating, moist and tender. Serve immediately, before the batter goes soggy.

The ideal accompaniments are, of course, chips (see p186), made first and held spread out on a baking tray in a warm oven, and tartare sauce (p43). What a lot of rather scary work. But only a select few chippies, along with some quite expensive restaurants, will make fish and chips as well as you can.

Grilled or baked fish

I treat them together because many recipes are interchangeable: you coat the fish with a marinade of some sort before putting it under or on the grill, or in the oven.

I was a bit sniffy about overhead grills in the Meat chapter (p270), but I like them for fish: they cook quickly enough to keep the fish tender, but not so fiercely that they scorch it. Even better is a barbecue. If you're going to use it a lot, you might invest in a fish-shaped, long-handled grilling basket, to which your fish won't stick as eagerly as it will to a grill rack.

You shouldn't go wrong if you bake fish at gas mark 6/200°C. Put it in a roasting tray or baking dish, and cover it with your marinade (see below). It's impossible to give accurate cooking times: the best I can manage is to tell you that a 500g fish may take 15 to 20 minutes. You'll have to stick a knife into the thickest part to see if the flesh is opaque and flaky.

Another way of baking fish is to enclose it in a loose parcel of foil. Lay it on the foil, sprinkle over some salt, pepper, herbs (chives, parsley or thyme, say) and a couple of tablespoons of white wine. Or (from Nigel Slater): for each fish, a couple of chopped chillies, a couple of smashed pieces of lemon grass, a couple of tablespoons of rice wine (or juice of a lime or 1/4 lemon), a few sprigs of coriander, some salt.

Bring up the sides of the foil to make a tent, and scrunch the edges together.

A grilled fish will take less time to cook than will a baked one. Line your grill pan with foil, place your fish in it, and cover with marinade. Turn it halfway through cooking; the operation requires some care.

2 marinades for a 500g grilled or baked fish

MARINADE 1

1 tbsp olive oil
1 garlic clove, finely chopped
1 shallot, finely chopped
1 dstsp herbs: parsley, chives, rosemary, thyme, oregano
Juice of 1/4 or 1/2 lemon, to taste
Salt

You can scatter these ingredients separately over the fish.

MARINADE 2

HOW TO MAKE IT

1 tsp coriander seeds
1 tsp cumin seeds
1 tbsp olive oil
1 garlic clove, finely chopped
1/2 tsp cayenne pepper
Salt

Put the coriander and cumin into a small saucepan, and place over a medium heat, shaking the pan until the spices give off a toasted aroma. Grind them in a mortar, or in a small electric mill or coffee grinder. Mix with the olive oil, garlic, cayenne pepper and salt, and pour over the fish.

FISH CAKES

Restaurant fish cakes come covered in golden, unbroken casings of breadcrumbs. I do not kid myself that I can reproduce that effect at home, satisfying myself merely with egg and flour.

HOW TO MAKE THEM

For 2
250g fish (any you like – smoked is good)
250g maincrop potatoes (King Edward, Maris Piper
    or Desirée, for example)
250ml milk
1 tbsp chopped parsley
1 tbsp chopped chives
1 egg, beaten
Flour on a plate
Oil and butter, for frying

Cook the fish (in the milk) and potatoes as described in the fish pie recipe (see p288), flaking the fish and mashing the potatoes. Stir the herbs, with salt (remember that smoked fish is salty), and pepper if you like, into the mash, with just enough of the egg (which will help to hold the cakes together) to make the mixture squidgy, but not enough to loosen it. Gently blend in the flaked fish.

Mould the fishy mash into 4 patties (or 2 large ones if you like; or 6 small ones). You won't be able to get them perfectly formed; not to worry. Dip them in the remaining egg, and roll them in the flour. Putting them in the fridge for half an hour will firm them up.

If you'd like breadcrumb-coated fish cakes, put the breadcrumbs on a separate plate. Roll the cakes in the flour first, then dip them in the egg, and then turn them in the breadcrumbs. It all gets a bit messy; and the fingers that touch the cakes come away with bits of bread stuck to them. But the browned coating may have a more coherent appearance than you expect.

Warm a frying pan over a medium heat; add enough oil (you choose what kind) and butter to coat the surface with a little to spare; when the butter is sizzling, slide in the fish cakes. (If the butter and oil mixture is not hot, the fish cakes will absorb more of it.) When you think the undersides are brown (probably after a couple of minutes), turn the cakes over gently. The other sides will brown faster.

It is possible that your fish cakes will not be properly warmed through by this time. Transfer them to a baking sheet, and put them into a gas mark 4/180°C oven for 10 minutes.

It would be a shame to waste the milk. Make a béchamel with it (see p46; with 250ml milk, you probably need about 25g each of butter and flour).

If you find you've used too much flour for the milk you have, you'll have to thin the sauce with more milk from the

fridge. Season with nutmeg and salt; pepper too, if you like.
Cook quickly 400g spinach (see p193); drain it in a colander,
pushing out the excess liquid with a wooden spoon. Stir the
spinach into the béchamel. You could chop it first.

Divide the sauced spinach between two warm plates, and
lay the fish cakes on top.

FISH SOUP, OR STEW

As Richard Olney puts it, inimitably: 'The line dividing a soup
from a stew is often infirm.' With both, you may get bowls full
of chunky bits of fish and vegetables, sitting in quite a lot of
liquid, and you need a spoon among your cutlery.

The following recipe will make a soup or a stew; if you
want a smooth soup, it will make that too.

HOW TO MAKE IT

For 8, easily
For the stock
Fish offcuts and heads (ask a fishmonger – the last time
    I did, I got charged £1 for a bagful)
Dark green parts of 3 leeks and of a bunch of spring onions,
    washed
Stalks and fronds of a fennel bulb
Stalks of a handful of flat-leaf parsley
2 celery sticks
8 garlic cloves

For the stew

1.5kg mixed fish,[1 – SEE WHY YOU DO IT] cut into chunks; if the fish
    are whole, use every bit of them, heads included
Olive oil
2 onions, roughly chopped (i.e., into chunks rather than
    fine pieces)
White and pale green parts of 3 leeks, sliced into
    fork-sized pieces
White and pale green parts of bunch of spring onions,
    roughly chopped
3 garlic cloves, chopped
1 fennel bulb, roughly chopped
4 celery sticks, roughly chopped
125ml white wine
2 bay leaves
500g carton passata
Zest of 1 orange
Salt
1.5 litres fish stock
450g new potatoes, boiled and sliced
Handful flat-leaf parsley, chopped
1 heaped tsp saffron fronds
Rouille (see below)
1 baguette, for croutons

Prepare the stock (for general remarks about fish stock, or
for a different recipe, which would also be fine here, see
p57). Cover the fish offcuts in cold water in a stock pot, bring
to a simmer, skim off the surface froth, and throw in the
vegetables. Cook on a very low heat for about 40 minutes.
Sieve, return the stock to a pan, and simmer until reduced to
about 1.5 litres.

Meanwhile, choose a large saucepan or casserole. I have a 28cm, oval Le Creuset; this stew filled it almost to the brim. Pour a layer of oil into the bottom, and throw in all the vegetables (except the cooked potatoes), cooking them over a low to medium heat until they start to turn golden. Pour in the wine, and allow it to bubble for a couple of minutes. Add the bay leaves, passata, orange zest, and salt to taste. Simmer until reduced and thickened.

Have your stock simmering in a saucepan. Turn up the heat under the stew, and pour in the stock. This is how you're supposed to make a bouillabaisse, liaising the stock and the vegetable mixture. You'll be surprised at how thick the sauce remains. Let it simmer for 5 minutes or so. Throw in the cooked potatoes, and bring back to a simmer. Check the seasoning.

Submerge the fish in the stew. Cook for 5 minutes longer. Check the fish: it should be ready. Turn off the heat, and stir in the parsley and saffron.

You could go on – after you've turned off the heat, but before you'd added the parsley and saffron – to make a smooth soup. It's one of my favourite things.

Place a food mill, with the coarse mesh attached, over another pan, and pass the stew/soup through it, pushing down on the fish with a wooden spoon and grinding as much of it through the disc as possible. Or: pour the soup into another pan through a colander, again pushing down on the ingredients with a wooden spoon.

Now refine the soup further: strain it into another pan (back into the first one, perhaps) through a sieve, pushing down on the pulp with the back of the ladle; discard the pulp that won't go through the mesh.[2]

Rouille is a fiery mayonnaise. Make a mayonnaise with

2 egg yolks, 300ml oil and 3 crushed garlic cloves (see p41); add to it 1/2 tsp cayenne pepper or 1 tsp harissa (p134). (Best add a little cayenne or harissa at first, and then taste before adding more.) You could put a little cayenne into the soup, too.

Cut the baguette into rounds, then into spoon-sized halves or quarters. Lay them on a baking tray, and put them in a gas mark 6/200°C oven until lightly toasted.

When you're ready to serve, add the saffron (with its soaking liquid if you've used threads), but perhaps not the parsley unless you particularly like it, and reheat the soup gently without boiling; check the seasoning. Place a bowl of croutons and a bowl of rouille on the table; guests dollop the rouille on to the croutons, which they float in their soup.

Soupe de poisson, as I just about dare to call this recipe, often comes with bowls of grated Gruyère to go with the croutons and rouille. As served in most restaurants, the cheese is sweaty and unappetizing, having been grated too far in advance. Even when it's in tip-top condition, I'm happy to leave it.

I don't dare to call the stew version of this dish a bouillabaisse, although it resembles it in composition. 'It is useless attempting to make a bouillabaisse away from the shores of the Mediterranean,' Elizabeth David warns.

WHY YOU DO IT

1 • What fish? I ask the fishmonger. The last time I made a soup, he gave me a snapper (which I chopped into three pieces), some coley and a cod fillet.

2 • Double straining. This process is fun, provided you're not in a hurry. The point is to get every last bit of liquid from the solid material.

A food mill will make the first round a lot easier than will a colander. Put the soup into it a portion at a time; use a wooden spoon to force the fish and vegetables between the metal attachment and the disc. Brief anti-clockwise turns of the handle can help the stuff to go through.

Sieve the pulp a portion at a time as well. The back of a ladle will help you to force the liquid through the mesh; throw away the dried pulp, and repeat.

MUSSELS

Mussels are enjoyable to cook: you get, as you lift the lid of the cooking vessel, the small miracle of the opened shells, as well as the wonderful briny, winey aroma. But preparing them is a little tedious.

HOW TO PREPARE THEM

Put them in a bowl of cold water. If any float, or are broken, throw them away. Drain them in a colander. Wash them again under running water, pulling off the beards if they have them. Look out for the following, and throw them away: any mussels that are open and do not close with a sharp tap on the side of the sink, or when you give them a little squeeze with your thumb and forefinger;[1 – SEE WHY YOU DO IT] any with shells that you can slide sideways with your thumb and forefinger (they will be gritty).

Phew. Now the more enjoyable bit.

HOW TO COOK THEM

As a starter, for 2 – double the quantities for a main course

500g mussels
1 tbsp olive oil
2 shallots or 1 onion, finely sliced
1 garlic clove, finely chopped
1 glass white wine
1 tbsp chopped parsley

Warm the oil over a medium heat in a heavy casserole or saucepan. Add the shallots or onion, and the garlic. Fry, stirring, for just a minute or 2, or for 3 or 4 if you're using onion; a slightly raw taste is fine in this context. Throw in the wine, and let it bubble (you shouldn't need salt – the mussel liquor is salty); tip in the mussels, and clamp on the lid. Shake the pan a few times while the mussels are cooking. Check after 3 minutes: a good many of them should be open. If not many are, they obviously need longer. If most are, put the lid back on for another minute to give the closed ones the benefit of the doubt;[2] if a few still refuse to cooperate, throw them away. (Don't try to prise them open to eat them – they'll probably poison you.) Scatter over the parsley.

Serve from their cooking vessel; eat the mussels from bowls, with spoons for the sauce, and bread to mop it up. If you're worried about grittiness in the sauce, scoop the mussels from the cooking vessel into a warm bowl, and strain the sauce over them through a sieve. Scatter on the parsley. You'll lose any shallot or onion you didn't pick up with the mussels, unfortunately.

You could add a couple of tablespoons of double cream to the cooking liquid at the end.

WHY YOU DO IT

1 • Alive, alive-o. Mussels, as you probably know, are alive – or should be – when you put them in the pot. If they don't close when you tap them, or open when you cook them, they are probably dead, not worth eating, and potentially harmful.

2 • Quick cooking. Mussels toughen, as do all protein-rich foods, if overcooked. It's a lot easier to deal with a moderate quantity of them: they should be ready at the same time. Crowded in a huge pile, they get varying access to the heat. When you cook more than a kilo of mussels, you may need to remove to a warm bowl the ones that open first; you'll protect them from toughening, and hasten the opening of the others.

Cakes and puddings sit a little oddly in this book. I've been trying to show you why, for the cook operating informally at home, recipes for most savoury dishes – certainly for most of the ones here – are merely templates, guides to the kinds of things that produce good results. We shouldn't need to get out a recipe book or the scales every time we make a pasta sauce or marinate a piece of chicken. If we know what works and why, we can save ourselves the effort and anxiety of feeling that we need to follow, to the letter, what the experts tell us to do.

However, in the world of puddings, recipes rule. A certain number of eggs will set a certain amount of liquid; a certain proportion of flour, eggs and liquid will make a batter. You cannot improvise your way around these determinants.

I take a recipe to be something that you attempt to follow with accuracy. In that sense, there haven't been many recipes in this book so far. But there are a few in the following pages.

I'm not saying that you have to obey them to the letter. You may prefer more sugar than I suggest, or less. I've left up to you the quantities you use of flavourings such as cinnamon. But I wouldn't advise you, for example, to be too liberal in your interpretation of the proportion of milk to eggs in the custard.

Sponge cakes

If a net is, according to Samuel Johnson, 'holes tied together with string', a cake may be described as bubbles contained by batter. You create the bubbles in three principal ways:

Beating (creaming) butter and sugar.

Incorporating a raising agent, either by using self-raising flour or by adding baking powder (and/or, in some recipes, bicarbonate of soda, an ingredient of baking powder).

Creating an egg foam.

The recipes here involve various permutations of these methods. But let's start with the technical stuff, some of which applies not only to cakes but to other sweet things in this chapter.

HOW/WHY YOU DO IT

1 • Equipment. Springform cake tins of 20cm and 23cm will cover a good many recipes. If you're making a sponge sandwich, you'll need two 20cm tins.

2 • Lining and greasing the tin. Place the cake tin on a piece of greaseproof paper, draw round it, and cut along the pencil mark. Smear a very small piece of butter on the base of the tin, stick the round piece of paper on top, and smear a little oil on the surface of the paper and round the sides of the tin. Oil works better than butter as a non-stick agent, because the solids in butter can be adhesive.

3 • Creaming. Generations of (mostly) schoolgirls suffered arm ache as they spent domestic science lessons – as they used to be known – mashing margarine, butter or some other shortening ingredient into sugar, and laboriously working away at the mixture until it lightened. These days, chefs tend to use hand-held mixers or food processors. Both machines require some manual intervention during the creaming process, because the mixture clogs up until it becomes properly amalgamated.

A creamed butter/sugar mixture, the texture of double cream, contains lots of air bubbles. It also separates the grains of flour, preventing lumps. This is why fats are known as 'shortenings': they interrupt the formation of gluten, which is a long chain of protein molecules.

4 • Sifting flour. Unnecessary, despite what recipes may say. You are unlikely to find weevils left behind in your sieve; and any airiness you give to the flour now will be lost when you stir it into the batter.

5 • Separating an egg. Crack the egg on the edge of a bowl, and allow the white to pour in. Gently, with your hand held over the bowl and your fingers straight, tip the yolk on to your fingers, opening them slightly to allow further white to slip through. Moving the yolk from hand to hand can encourage this process. (I got this technique from the opening sequence of a TV biopic of Elizabeth David.)

6 • Whisking whites. Try to avoid letting any trace of yolk creep into the egg white. Don't add salt or, despite what some experts recommend, lemon juice or vinegar: they soften egg whites.

I use a hand-held whisk, feeling that, in spite of the work involved, it enables me more accurately to judge the progress of the foam.

Use a large bowl, and tip it towards you, so that the whisk gets access to as much egg white as possible. You should stop beating when the whisk, lifted from the foam, creates peaks, which do not subside. It's tempting to carry on, just to make sure you've got the right consistency. Resist. Further beating causes the peak stage rapidly to be succeeded by collapse.

7 • Dropping the cake. In *The Science of Cooking*, Peter Barham offers the bizarre recommendation that, on removing the cake from the oven, you drop it from a height of about 30cm on to a hard surface. (An average ruler is about 30cm long.) The theory is this: as a cake cools, the air bubbles in it deflate, like collapsing balloons. Dropping the cake allows some of the bubbles to break, letting in air, which sustains the structure.

8 • Turning out. Some recipes recommend that you wait 5 minutes, just long enough for the cake to contract from the side of the tin, before turning it out on to a wire rack. If you do this, place the cake top-side (the firmer one) down. I find that bits of cake stick to the rack anyway, and I usually allow the cake to cool in the tin.

9 • Zesting. I scrape the lemon against the finest mesh on my cheese grater, and try to grate only the skin: the white pith underneath is bitter.

BASIC SPONGE

This sponge does contain a certain amount of gluten, which you develop when you blend the flour with the sugar and butter, and when you stir in the eggs. If you have ever handled bread dough, which has a lot of gluten, you'll know that it has an elastic quality. Here, that elasticity maintains the structure of the cake as the air bubbles expand. If the structure were to break and the bubbles to pop, the cake would collapse. The texture of the cake is, nevertheless, foamy.

HOW TO MAKE IT

100g self-raising flour (or plain flour, plus 1 tsp baking powder)
100g caster sugar
100g softened butter
1 tsp vanilla essence (optional)
2 eggs, beaten

Preheat your oven to gas mark 4/180°C. Put in a baking sheet.

Line and grease a 20cm springform cake tin.[2 – SEE HOW/WHY YOU DO IT, P310]

In a food processor, blend the flour, sugar and butter, in short pulses, until you have a stodgy mass. Tip the mixture into a bowl, and stir in the vanilla (if using – I like it, but you may prefer less, or none at all) and a portion of the egg. Keep adding egg until you have a gloopy batter; it should drop off a spoon, but reluctantly. Don't feel obliged to use all the egg; but, if you have done so before you get to the gloopy stage, add a little milk too.

Tip the batter into the cake tin, spread it out and level the surface, and put the tin into the oven on top of the baking sheet, which helps to convey the heat. Bake for about 25 minutes, or until an inserted skewer emerges clean.

Drop the cake tin from a height of about 30cm on to a hard surface (I hope the spring is secure).[7] Allow the cake to cool before turning it out.[8] I keep it in greaseproof paper, wrapped inside foil.

Add the zest of a lemon, or of an orange, to the mix. And/or poppy seeds. Or any other dry flavourings.

Make a sandwich: Double the ingredients, and divide the batter between two tins, baking the layers side by side.

Fillings. You might go for a cream tea theme, with a layer of jam topped by whipped cream. Or try lemon curd, from *Geraldine Holt's Cakes*: 1 lemon; 60g caster sugar; 1/4 tsp cornflour; 1 egg; 30g butter. Use a small bowl that will rest in a saucepan of simmering water without the base touching the water's surface. Grate in the lemon zest, [9 – SEE HOW/WHY YOU DO IT, P312] and add the juice too, along with the sugar, cornflour and egg. Cook the mixture, stirring all the time, above the simmering water. After about 5 to 7 minutes, the mixture should start to thicken. Remove the bowl, and stand it in cold water; while the mixture still has some warmth, stir in the butter in small pieces. Allow the curd to cool before spreading it on the cake.

VICTORIA SPONGE

The same ingredients as above, but, you'll notice, in a different order. Creaming the butter and sugar before adding the egg and gently folding in the flour results in a cake that is lighter in gluten, and that therefore has a lighter texture. The drawback is that the air bubbles are more likely to burst in the less elastic batter, particularly if your mixture is too loose.

100g caster sugar

100g softened butter

2 eggs, beaten

1 tsp vanilla essence (optional)

100g self-raising flour (or plain flour, plus 1 tsp baking powder)

Preheat your oven to gas mark 4/180°C. Put in a baking sheet. Line and grease a 20cm springform cake tin.[2] – SEE HOW/WHY YOU DO IT, P310

Cream the sugar and butter until the mixture is soft and fluffy.[3] Beat in about three-quarters of the egg – this may be enough. Add the vanilla, if you're using it, and gently fold in the flour. You should have a gloopy batter, which will drop off a spoon, but reluctantly. If the batter is too stiff, gently add more egg. If it's still too stiff, add a little milk.

Tip the batter into the cake tin, spread it out and level the surface, and put the tin into the oven on top of the baking sheet, which helps to convey the heat. Bake for about 25 minutes, or until an inserted skewer emerges clean.

Drop the cake tin from a height of about 30cm on to a hard surface.[7] Allow the cake to cool before turning it out.[8] I keep it in greaseproof paper, wrapped inside foil.

VARIATIONS

See the suggestions for the basic sponge recipe, p314 Here's an extra one: lemon drizzle. Put the zest [9] – SEE HOW/WHY YOU DO IT, P312 and juice of 2 lemons in a bowl, and stir in 100g granulated sugar until dissolved. When you remove the cake from the oven, spoon this syrup all over it.

GENOESE SPONGE

In this sponge, the eggs produce the foam – there is no other raising agent. A cake the size of the sponges described above would again have 100g flour and 100g sugar, but with 4 eggs, and 50g melted butter. You need an electric whisk to beat together the sugar and eggs; eventually, after a good 10 minutes and maybe more, they increase in volume by as much as six times, and become stiff and very pale. Then you fold in the flour. Last (because the fat collapses the air bubbles), you stir in the butter, quickly transfer the batter to the tin, and bake it as in the recipes above.

CHOCOLATE CAKE

An alternative and easier way of creating an egg foam is to separate the eggs and beat the whites only. This cake is an Elizabeth David recipe, and is seductively moist and gooey. I recommend that you eat it all on the day you make it: by day two, it becomes somewhat compacted.

HOW TO MAKE IT

84g butter
100g caster sugar
110g dark chocolate
55g flour
3 eggs, separated[5] – SEE HOW/WHY YOU DO IT, P311

Elizabeth David recommends you use a 1 1/2 pint loaf tin. I have no idea how large that is. Mine is 7cm x 16.5cm, and seems to be roughly the right size. Line it and grease it.[2 – SEE HOW/WHY YOU DO IT, P310] Heat the oven to gas mark 4/180°C.

Cream the butter and sugar.[3] (Elizabeth David does not include this instruction; but I have found that it produces a lighter cake.)

Cut up the chocolate, and melt it in a bowl suspended above a pan of barely simmering water. This, I find, is the easiest way of ensuring you do not overheat the chocolate and turn it grainy. Remove it from the heat as soon as it melts, and stir into it the creamed butter and sugar, flour and egg yolks. (Or, if the chocolate bowl is not large enough, pour the chocolate into the bowl with the butter and sugar – but you're bound to leave some chocolate behind.)

Beat the egg whites until they form soft peaks.[6] Fold them into the chocolate mixture, gently turning and lifting until the mixture is amalgamated, but without beating the air bubbles from it.

Pour the mixture into the tin, and bake on a baking sheet for 35 minutes, or until an inserted skewer emerges clean.

Drop the cake tin from a height of about 30cm on to a hard surface.[7] Allow the cake to cool before turning it out.[8] You could keep it in greaseproof paper, wrapped inside foil; but, as I suggest above, you may prefer to eat it right away.

2 cakes without flour

I have never succeeded in making a satisfactory sponge by substituting gluten-free equivalents in the above recipes. Gluten-free flour and xanthan gum (which supplies the binding qualities of gluten) produce cakes with powdery textures, I find. Other cooks, specialists in this field, use combinations of flours to come closer to the ideal, more crumbly sponge; but if I want a gluten-free cake, I prefer to turn to recipes that work in their own right.

ALMOND CAKE

HOW TO MAKE IT

This is a slight adaptation of a recipe by Hugh Fearnley-Whittingstall. The chief difference is that I prefer to use two lemons rather than one – the flavour can seem a little dry otherwise.

8 egg yolks[5] – SEE HOW/WHY YOU DO IT, P311

250g caster sugar

1 tsp vanilla essence

Zest of 2 lemons[9]

1 tsp ground cinnamon

300g ground almonds

6 egg whites[5]

Icing sugar

Heat the oven to gas mark 4/180°C, and put in a baking sheet. Line and grease a 20cm springform cake tin.[2] – SEE HOW/WHY YOU DO IT, P310

Whisk together the egg yolks and sugar until light and creamy (see Genoese sponge, p316). Stir in the vanilla, lemon zest and cinnamon. Fold in the ground almonds.

Beat the egg whites until they form soft peaks.[6] Add about a third of them to lighten the batter. Now you can fold in the remainder of the whites without having to thrash the air out of them.

Scrape the batter into the cake tin. Bake, on the baking sheet (which helps to convey the heat), for 40 to 45 minutes, or until an inserted skewer comes out clean.

Drop the cake tin from a height of about 30cm on to a hard surface.[7] Allow the cake to cool before turning it out.[8] Dust it with icing sugar, if you like.

LEMON POLENTA CAKE

This cake is given a lift by baking powder rather than by egg foam. Gluten-free baking powders are available.

HOW TO MAKE IT

225g butter
225g caster sugar
3 eggs
200g ground almonds
125g quick-cook polenta
1 tsp baking powder
1 tsp vanilla essence
2 lemons – zest and juice[9] – SEE HOW/WHY YOU DO IT, P312

Set the oven to gas mark 4/180°C. Line and grease a 20cm springform cake tin.[2] – SEE HOW/WHY YOU DO IT, P310

Cream the butter and sugar.[3] Add the eggs, and stir in the almonds, polenta, baking powder, vanilla and lemon.

Scrape this batter into the tin. Bake, on the baking sheet (which helps to convey the heat), for 40 to 45 minutes, or until an inserted skewer comes out clean.

It is especially important to let this cake cool before you turn it out. It will be quite crumbly until then.

Custard puddings

CUSTARD

As an accompaniment to puddings for 4 to 6 people
1/2 vanilla pod, or 1 tsp vanilla essence
150ml milk
150ml double cream
3 egg yolks[1] – SEE WHY YOU DO IT
2 tbsp, or about 30g, caster sugar[2]

Slit the vanilla pod lengthways with a sharp knife; scrape the seeds into the milk and cream. Warm the milk and cream with the pod and seeds in a saucepan until bubbles appear; turn off the heat, cover the pan, and leave for 20 minutes.[3] Lift out the pod. Return the pan to a simmer.

Or, if using vanilla essence: bring the milk/cream and vanilla essence to a simmer.

Beat the egg yolks and the sugar. Pour a little hot milk into the mixture, whisking;[4] then a little more; then tip this mixture back into the saucepan. Over the gentlest heat, stir the custard until it starts to thicken; remove it from the heat immediately, and pour into a warm, but not hot, jug.[5]

Serve right away, or keep the custard covered in the fridge, where it will thicken, to serve cold.

WHY YOU DO IT

1 • Yolks only. They will produce a richer custard than would whole eggs. You'll need three yolks, at least, to set as much custard as you would with two whole eggs. For advice on separating eggs, see point 5, p311.

2 • How much sugar? Recipes vary wildly. Of course, it's a matter of taste.

3 • Infusion time. To let the vanilla flavour the milk and cream. If you want a milder vanilla flavour, without the seeds in the milk, drop in the whole pod without slitting it; you'll probably be able to use it again. If you don't want vanilla seeds in your custard, pour the milk and cream – after the infusion process – into another pan through a sieve.

4 • Milk into eggs. Put the hot liquid into the cold liquid, so that the volume of the cold liquid is greater, and whisk all the time to distribute the heat. That way, you avoid making runny scrambled egg rather than custard. If you put the eggs into the simmering milk, they will get a sudden exposure to boiling heat, and curdle. (Similar considerations – along with ones of social status – inform the decision about whether to put milk or tea into your cup first.)

5 • Don't boil. Your safest bet is to warm the custard in a double boiler above simmering water. Or put it into a bowl – a glass one will do – that fits inside a saucepan. The bowl shouldn't touch the water. But you can get away with stirring the custard vigorously in a heavy-bottomed pan above a very gentle flame, perhaps with a heat disperser between flame and pan. To be thorough, clean any scorched milk from the bottom of the pan before returning the custard to it.

If bits of custard stick to the bottom of the pan, they will curdle – the egg will scramble, in other words. The egg will also scramble if the liquid approaches boiling point. Heat the custard gently, and stop heating it, despite the temptation to get it just a bit thicker, as soon as it has the consistency of a pouring cream.

BAKED CUSTARD

HOW TO MAKE IT

For 6
Regular version[1] – SEE WHY YOU DO IT
4 whole eggs
300ml milk
300ml double cream

Rich version[1] – SEE WHY YOU DO IT

6 egg yolks

150ml milk

450ml double cream

For both versions

1 vanilla pod, split lengthways, or 2 tsp vanilla essence

50g caster sugar, or 1 tbsp (about 50g) honey

Make the custard as you would the pouring version, above. (If you use honey, you'll be able to beat it with the eggs – thick honey will go runny.) Pour it into an ovenproof dish, lightly greased with oil, and put the dish inside a pan – a roasting pan, say. Pour hot water into the pan to come halfway up the sides of the custard dish.[2] Bake at gas mark 3/160°C for about 35 to 45 minutes, or until the custard is set. Serve hot or cold.

WHY YOU DO IT

1 • How rich is your taste? I sometimes prefer the lighter, less unctuous texture of the custard made with whole eggs.

More egg yolks, and more cream, are required to set the richer version. Of course, you could perm these ingredients: use 2 whole eggs and 3 yolks, for example; or, in the yolk-only version, use less cream and another egg yolk (making 7).

2 • The bain-marie. The temperature of your oven is 160°C – high enough to curdle the custard in contact with the sides of the cooking vessel. The water in the bain-marie – the bath in which the cooking vessel sits – cannot of course get hotter than 100°C, and protects the custard from the oven heat. Don't allow the water to come too high up the sides of the custard dish, or it might bubble over.

PANETTONE BREAD AND BUTTER PUDDING

HOW TO MAKE IT

For 6

6 slices panettone[1] – SEE WHY YOU DO IT

Butter

Ingredients for baked custard (see above)

Lightly oil an oven dish. Butter the panettone slices (on one side only, if you want to make a token concession to heart-preservation), and place them in the dish, as evenly as possible.

Make the custard according to the baked custard recipe, and pour it over the panettone. It should cover the slices, but not swamp them; add another slice or two of cake if you think there's too much liquid.

On the other hand, you might find that the panettone is quite bulky; if so, start with fewer slices, pour on the custard, and then add more cake if there's room.

Place in a bain-marie (see Baked custard, p323), and bake in a gas mark 3/160°C oven for 45 minutes, or until the custard is set.

VARIATIONS

This recipe is itself a variation, on traditional bread and butter pudding made with sliced white bread. One reason why I like it is that the cake provides the fruit and spice that you would otherwise have to supply yourself.

For the traditional version, use about eight thin slices, crusts removed, of white loaf, again buttered on one side. If you want fruit, scatter about 20g each of raisins and sultanas over the bread slices; you could soak them first for 20 minutes in a few teaspoons of brandy or liqueur (pouring the alcohol over the bread with the fruit). If you want spice, flavour the custard with a pinch or two of grated nutmeg and cinnamon.

WHY YOU DO IT

1 • **How much cake?** As I imply in the recipe, I cannot give precise quantities. I don't know how big your panettone is, or how thickly you've cut your slices. What I do know is that most bread and butter pudding recipes give too high a proportion of bread for my taste: I think that bread and butter pudding should consist of bread (or cake) with a liberal lubrication of custard, rather than simply of eggy bread. For this reason, I don't leave the bread (or cake) to soak up the custard before baking it, as some recipes advise. It absorbs plenty of liquid while baking, anyway.

RICE PUDDING

This is a Heston Blumenthal recipe – rice pudding meets risotto. I've reduced the quantity of milk he gives, because I didn't need so much. Perhaps your rice will absorb more; you can add it during cooking.

For 6
600ml milk
1 vanilla pod, or 2 tsp vanilla essence
135g arborio or carnaroli rice
110g caster sugar
1 pinch nutmeg
150ml double cream
2 egg yolks

Split the vanilla pod lengthways; put it, with scraped-out seeds, in the milk, bring the milk to a simmer, cover, and turn off the heat. Leave for 20 minutes.

If you don't want the seeds, sieve the milk into another pan.

Bring a pan of water to the boil, throw in the rice, bring back to the boil, simmer for 3 minutes, and drain.
1 – SEE WHY YOU DO IT

Add the sugar and nutmeg to the milk, and return to a simmer (if you're using vanilla essence, you'll have started here). Add the rice and simmer it, stirring. It should take 20 to 30 minutes to become tender.

If the rice is still drowned in liquid, strain some milk into another pan and simmer it until it reduces, before returning it to the main dish. A certain amount of runniness is fine, particularly if you're going to serve the pudding cold: it will thicken as it cools.

Add the double cream; cook for a couple of minutes longer. Remove from the heat, add the egg yolks, and stir. Blumenthal tells you to stir the pudding for 3 minutes; but you can probably get away with being a little less assiduous.

Serve hot or cold.

WHY YOU DO IT

1 • Blanching the rice. To blanch is to part-cook something in boiling water. In this case, you do it to lessen the starchiness of the rice.

LEMON SURPRISE PUDDING

This is a lemony sponge on top of a lemony custard. Nigel Slater, with credit to Margaret Costa's classic *Four Seasons Cookery Book*, reproduces it in *Real Cooking*, but includes an egg too few and a lemon too many for my taste: I find the custard too thin, and sharp.

In *Appetite*, he offers an updated version, with orange

and lemon, and gives the same number of eggs but less milk and flour.

The following recipe is halfway between the two.

HOW TO MAKE IT

For 6

100g butter

175g caster sugar

Yolks of 5 (or 6, for a thicker custard) eggs, whites
     of 4[1] – SEE HOW YOU DO IT

2 lemons – zest and juice[2]

500ml milk

50g plain flour

Cream the butter and sugar.[3] Beat in the egg yolks, and add the lemon zest and juice. The mixture won't look too good, but don't worry: it's going to look even worse in a minute. Pour in the milk, and gradually whisk in the flour (see the batter recipe, p225).

Beat the egg whites with a balloon whisk – use an electric beater if you prefer (see p311) until you can make snowy peaks with them.[4] Fold them gently into the batter. Pour the mixture into a buttered oven dish; it should come halfway up the sides.

Put the dish into a roasting pan or other vessel, and pour in hot water to come halfway up the sides of the pudding dish (see The bain-marie, p323). Bake in a gas mark 4/180°C oven for about 50 minutes, or until the sponge is golden and, well, spongy.

You could try an orange surprise pudding, with the zest and juice of 2 oranges; or use 1 orange and 1 lemon.

HOW YOU DO IT

1 • Separating. See point 5, p311.

2 • Zesting. See point 9, p312.

3 • Creaming. See point 3, p310.

4 • Whisking. See point 6, p311.

PANNA COTTA (POSH BLANCMANGE)

HOW TO MAKE IT

For 6

3 leaves gelatine, or 3 tsp powdered
    gelatine[1] – SEE WHY YOU DO IT

450ml double cream

150ml milk

60g caster sugar

2 vanilla pods or 1 tsp vanilla essence

Put the gelatine (leaves or powder) in a small saucepan, and sprinkle over just enough water to soak it. Leave for 10 minutes.

Meanwhile, pour the cream and milk into a small saucepan, and add the sugar. Split the vanilla pods from end to end, scrape the seeds into the milk and cream, and throw in the pods as well; or simply add the vanilla essence. Bring the pan to a simmer, and allow the contents to bubble gently for 5 minutes.

Pour the cream and milk mixture – through a sieve if it contains vanilla pods and seeds – into a jug.

Put the saucepan with the gelatine on to a ring on the hob at its lowest setting. As soon as the gelatine turns watery, take it off the hob and stir until it completely dissolves – overheating disables its thickening qualities. (Another way with leaf gelatine is to soak it in cold water for 5 minutes, squeeze it gently, and then dissolve it in the warm panna cotta mixture.)

Allow the cream and milk to cool, and stir in the gelatine, gently but thoroughly. Pour the mixture into 6 ramekins, cover with clingfilm, and chill for at least 3 hours.

If you want to turn out the panna cottas, grease the ramekins with a little oil first. Dipping the chilled ramekins briefly in hot water will encourage the panna cottas to contract from the sides.

VARIATIONS

Dark or white chocolate panna cotta. Chop 100g chocolate into small pieces. Allow the cream/milk to cool slightly when you take it off the heat, and then stir in the chocolate until it dissolves. Or add cocoa powder – about 3 tbsp. With white chocolate, you may not need any more sugar. How much sugar you add to the dark chocolate mixture is a matter of taste – no more than 30g, I'd suggest.

Lemon panna cotta. Add the zest and juice of a lemon (see p312) to the cream/mixture when you take it off the hob. The extra liquid should not be enough to stop the mixture setting.

A little **fruity acidity** offsets panna cotta nicely. Serve with strawberries (sliced), raspberries, blackberries, or other berries. Or try a simple blueberry compote: empty a punnet of blueberries into a saucepan, and heat with a dessertspoon of caster sugar until they burst. Chill. Serve a spoonful of compote with each panna cotta.

WHY YOU DO IT

1 • Gelatine. These quantities have worked for me. But leaf gelatine in particular can vary in setting qualities. Check the packet instructions.

Fruit puddings

FRUITS IN SYRUP

Try this ratio: 1g sugar – it doesn't matter whether it's caster or granulated, but remember that caster sugar will be sweeter by volume – to each 2.5ml water. So your syrup might be

100g sugar and 250ml water; put them in a pan, and bring to the boil, to dissolve the sugar. You might add the following flavourings: a cinnamon stick or pinch of cinnamon; a pinch of nutmeg; a vanilla pod or teaspoon of vanilla essence.

You can poach any number of fruits in this syrup. For example: apricots, halved and stoned; plums, likewise; peaches; nectarines; strawberries; raspberries; blackberries; blackcurrants; any more you can think of. Put them into the pan with the hot syrup and simmer gently, until tender. Use a pan that's wide enough to allow all the fruit to get a bath. If you need more liquid, simply add more water and sugar; but 250ml should be plenty for about 600g fruit – you don't need to drown it, and you can use just enough syrup to come halfway up the fruit, which you turn halfway through cooking. Cover the pan; but make sure that the contents don't boil too vigorously, or catch and burn on the bottom.

When the fruit is tender, remove it with a slotted spoon. (Plums, turned halfway through cooking, might be ready in no more than 6 minutes. Blackberries, which should acquire a melting texture, will take longer.) Turn up the heat under the pan to reduce the liquid to a syrupy consistency. Pour it back over the fruit.

Serve hot or cold, with cream or custard (see p320).

Rhubarb

It's not a fruit, of course, but a vegetable – a leaf stalk. But we eat it sweetened, as a pudding. You can poach it, as above; but I prefer to cook it in a minimal amount of liquid. Wash the stems, and cut them into short (3cm, say) pieces. Melt enough butter in a saucepan to coat the bottom; add the damp rhubarb, and cook it, with the pan covered, over a gentle heat. The rhubarb will throw off liquid. Now you can turn up the heat a little, and add sugar. You can uncover the pan too, because you want the liquid to evaporate, and because with stirring you can give all the rhubarb access to the heat.

Simmer until tender. Tip the rhubarb into a sieve, allowing the liquid to pour into another pan; boil the liquid to thicken it. Re-combine it with the rhubarb; check for sweetness, adding more sugar if necessary. Serve hot or cold, with cream or custard (see p320). Or use it in a crumble (see below).

Forced rhubarb – the slim, delicately pink stalks you get early in the year – respond particularly well to baking, which has the advantage of keeping the stalks intact. Cut the rhubarb into fork-sized pieces, place in an oven dish, scatter over the sugar, and bake at gas mark 6/200°C for 20 to 30 minutes, or until tender. Stir everything from time to time. You might include cinnamon, nutmeg, vanilla essence, orange juice, or other flavourings. If you have too much liquid, remove it to a pan, boil it until it turns syrupy, and pour it back over the rhubarb.

Sorry to be vague about the quantity of sugar. Say you have 600g rhubarb (by the way, get rid of the leaves, which are reputed to be poisonous): add 2 tbsp sugar. It probably won't be enough. But it's better to start with too little, and to add more as your taste dictates, than to start with too much, when your only way of tempering the sweetness will be to add more rhubarb.

Fruit fools

Stew rhubarb or gooseberries as in the basic rhubarb recipe (above), but without the butter. Drain off the liquid into a pan; put the rhubarb or gooseberries into a sieve held over the same pan, and crush them with a wooden spoon. Boil all the liquid in the pan until it is syrupy; re-combine it with the rhubarb or gooseberries. Check for sweetness. Put the fruit into the fridge, to acquire a holding, slushy consistency.

You can give the same treatment to redcurrants and blackcurrants.

About 600g fruit (for stoned fruit, that's the weight when the stones are removed) should be good for 6 people. How much cream you use is a matter of taste; I suggest a similar volume to that of the fruit – 400ml, say. Pour the cream – double or whipping – into a bowl, and whisk with a balloon whisk, until it thickens but before it goes firm. It should have a coating, but still liquid, consistency. Combine it with the fruit, and chill again.

Strawberries and raspberries don't want cooking. Pour sugar on them, bash them up with a fork, check the sweetness, and combine with the thickened cream. Chill.

FRUIT CRUMBLE

HOW TO MAKE IT

For 6
125g butter
200g plain flour
75g caster or demerara sugar

More sugar for the fruit

600g unstoned fruit (if you're using plums or damsons, you'll probably need about 1kg; if you're using apples such as Bramleys, about 900g)

This is one job for which I always use a food processor. Cut the butter into pieces, and whizz them with the flour briefly; the mixture should resemble fine breadcrumbs. Then stir in the sugar.

Or: work the butter into the flour with your fingertips. Some people find it satisfying.

Apple crumble. Most crumble recipes underestimate the amount of softening you need to give to the fruit before you bury it in crumble. If fruit goes into the oven crunchy, it will probably emerge crunchy.

Peel the apples, halve and quarter them, take out the cores, and cut the flesh into wedges. You can stop them going brown if you drop them in water into which you've squeezed some lemon juice. In a heavy saucepan, melt enough butter to coat the pan bottom, add the apples (damp from their soaking), and cook them gently, with the pan covered.

When there's a bit of liquid in the pan, you can add some sugar – start with a tablespoon, and then taste – without fear of caramelizing it.

You may find that your apples give off a good deal of liquid; uncover the pan to allow it to evaporate, and continue to cook, stirring regularly. But others may need the help of a little additional water. They'll take 15 to 20 minutes to soften. You could add a little cinnamon and nutmeg.

If you can't find Bramleys, try other varieties with a bit of acidic character to them: Cox's, for example.

Tip the cooked apples, having checked that they're sweet enough, into a pie dish or other ovenproof dish. Cover with the crumble. The quantity of flour, butter and sugar given above is only a guide, and may be over-generous; don't use it all if it threatens to overwhelm the fruit. You want a light, buttery layer of crunchy crumbs, not a thick, dry layer of compacted stodge.

Put into a gas mark 5/190°C oven for about 30 minutes, or until the top of the crumble is golden.

Apple and blackberry is good (in a 600g to 300g ratio, say). Poach the blackberries (see p331), and add them, with a reduced and thickened syrup, to the cooked apples.

Or **apple and rhubarb**; again, cooked separately (see Rhubarb, p332). Try equal volumes of each fruit, weighing about 900g in total.

Plum crumble. Cut 1kg of plums – or more if you like – in half, and remove the stones. Simmer in syrup (see p331), in a wide, covered pan, for about 2 minutes on each side; they should just be starting to tenderize.

Transfer them in a slotted spoon to an oven dish; reduce the syrup to a thick liquid, and pour it over the fruit. Cover with the crumble, and bake as above. You don't need much syrup: just enough, say, to come a third of the way up the stoned and halved fruit.

Blackcurrant and/or redcurrant crumble. Simply mix 600g of them in an oven dish with 3 tbsp sugar, cover with crumble, and bake.

PLUM CLAFOUTIS

Is it a custard? Is it a batter? Recipes disagree. The one I like best, of which the following is an adaptation, is quite a custardy one; it comes from Raymond Blanc's *Blanc Mange*. My version never resembles the one on the cover of the book – but then, you don't expect that to happen, do you? It tastes pretty good, nonetheless.

HOW TO MAKE IT

For 6
Syrup: 150ml water, 50g caster sugar
12 plums, halved, stones removed
1/2 vanilla pod or 1 tsp vanilla essence
55ml milk
55ml double or whipping cream
3 eggs
90g caster sugar
200g plain flour

In a wide saucepan, warm the syrup; when it's simmering, add the plums (which should all have access to the water), and cook for 4 to 5 minutes, turning them halfway through. You don't need to put the lid on the pan. Check to see that the plums are soft, but not mushy. You can let them cool in the saucepan, covered.

Split the vanilla pod lengthways, add it, with scraped-out seeds, to the milk and cream in a saucepan, and bring to a simmer. Turn off the heat, cover the pan, and let the vanilla flavour infuse for 20 minutes. Remove the pod.

You want to get air into the eggs. Whisk them with the sugar using a hand-held whisk until they have expanded considerably in volume, and become very pale. It could take 10 minutes, or longer. Stir in the flour, and then the milk and cream, which – if you want to lose the vanilla seeds – you can pour on to the mixture through a sieve. (If you're using vanilla essence, you won't need to have heated the milk and cream beforehand. Simply add the vanilla now.)

Or you could introduce airiness by separating the eggs and whisking the whites (see points 5 and 6, p311). Pour the milk and cream – in which, this time, you have dissolved the sugar – gradually into the yolks, stirring all the time; gradually stir in the flour, and then fold in the whites.

I use a flan dish with a diameter of about 23cm. Grease it with a little oil; lay the plums in it; pour over the batter; bake in a gas mark 4/180°C oven, for about 30 to 35 minutes, until just set.

You could boil down the syrup to use as a sauce.

You're supposed to wait until the clafoutis is tepid; but I don't see why you shouldn't eat it hot.

VARIATIONS

Raymond Blanc's recipe is for apricot clafoutis; the apricots are prepared as above. He also includes some alcohol: for the recipe above, you'd want about a table-spoon of amaretto liqueur, poured into the custard/batter mixture. A classic clafoutis involves cherries, which don't require pre-cooking and are simply mixed with the custard/

batter. Guests have to pick out the stones themselves, while eating. How many would you like? About 36 in total, I'd suggest.

You might prefer the simpler (and less floury) clafoutis batter from Arabella Boxer's *Mediterranean Cookbook*: 3 eggs, 3 tbsp caster sugar, 3 tbsp flour, 600ml heated milk. The technique is the same.

Two mousses

CHOCOLATE MOUSSE

HOW TO MAKE IT

For 6

150g dark chocolate (with at least 70 per cent cocoa solids –
    Green & Black's makes a fine mousse)
25g butter, cut into small cubes
6 eggs, separated[1] – SEE WHY YOU DO IT

Break the chocolate into its squares, and melt it in a bowl held above a pan of simmering water. (I have a Pyrex bowl that rests on the edges of a saucepan.) Stir the chocolate to encourage melting, and remove it from the heat as soon as, or slightly before, all the lumps have disappeared.[2] Drop in the butter.

Beat the egg whites until they form soft peaks.[3] Take about a quarter of the white and beat it into the chocolate and butter. The mixture should retain the texture and consistency of thick chocolate sauce. Now stir in the yolks.[4]

Pour the chocolate mixture over the egg whites. Fold it in, using a lifting and turning motion with the spoon until amalgamated.[5] Transfer to a dish, or to individual ramekins, and refrigerate for at least 3 hours.

VARIATIONS

You could include double or whipping cream (150ml, say), whipped until it is thick but not until it stiffens. I find a mousse that's eggy, chocolatey and creamy as well a bit much. But, if you're nervous about raw eggs, you could go for simple chocolate pots: mix 150g melted chocolate with 300g whipped cream. Refrigerate. That's it.

Chocolate orange mousse: introduce the grated zest of an orange. This mousse is meant to be bitter. I think that it would be a shame to sweeten it.

WHY YOU DO IT

1 • Separating eggs. See point 5, p311.

2 • Melt the chocolate, don't cook it. If the chocolate becomes too hot, it becomes grainy. This method of melting it is the easiest to control, I find.

3 • Whisking whites. See point 6, p311.

4 • Beating in a portion of egg white helps to loosen the chocolate/butter mixture, so that you can fold in the rest of the white gently (see below). It cools the mixture, too. If the chocolate is hot when you add the egg yolks, it will seize up, and the mousse will not be light.

5 • You're trying to maintain the foam, which would be destroyed if you merged the white and the egg/chocolate/butter mixture too vigorously.

LEMON MOUSSE

This mousse is gorgeous. It has a delicious balance of citric acidity and sweetness, as well as a lovely, foamy texture.

The egg white cannot sustain the foam on its own, because of the lemon juice, and needs the assistance of gelatine. Tricky stuff, gelatine. The trick here – not one about which recipes are very helpful – is to blend the gelatinous mixture and the egg white at the right moment. Do it too soon, and the mixture separates and sinks; too late, and the mixture is too well set to be blended.

HOW TO MAKE IT

For 4
2 tsp gelatine, or 2 sheets of leaf gelatine[1] – SEE HOW YOU DO IT
3 eggs, separated[2]
150g caster sugar
2 lemons, juice and grated zest
125ml double cream

Put the gelatine (leaves or powder) in a small saucepan, and sprinkle over just enough water to soak it. Leave for 10 minutes. Meanwhile, combine the egg yolks and sugar in a bowl, and beat them with a wooden spoon until they turn pale yellow. Beat in the lemon juice and zest.

Put the saucepan with the gelatine on to a ring on the hob at its lowest setting. As soon as the gelatine turns watery, take it off the hob and stir until it completely dissolves – overheating disables its thickening qualities. Add it to the egg and lemon mixture, stirring gently but thoroughly.

You leave this mixture until it starts to set. How long is this? Recipes tend to sidestep the question. I find that after an hour in the fridge, the mixture still swirls in the bowl, but is no longer runny. This seems to be the moment one wants.

Whisk the cream until it thickens, but stop before it becomes stiff – the transition is rapid, so take care. In a separate bowl, and with a separate – or at least clean – whisk beat the egg whites until they form soft peaks.[3] Fold into the whites the cream and the egg yolk mixture, which should have the consistency of a collapsing jelly. Again, perform the action gently, but do so until the mousse is thoroughly blended.

Spoon the mousse into a bowl, cover with cling film, and refrigerate for at least 6 hours.

HOW YOU DO IT

1 • **Gelatine**. For advice about quantities, see p331.

2 • **Separating.** See point 5, p311.

3 • **Whisking the whites**. See point 6, p311.

Cheesecakes

My favourite pudding. Hence my self-indulgence in giving several recipes here.

It must have a good base, one with a crunch to contrast with the smooth, often tangy filling. The proportion of butter I give – higher than in many recipes – helps to bind the crumbs. Even so, you may find that the base of a baked cheesecake is more crumbly than one that is simply chilled.

HOW TO MAKE IT

THE BASE

150g digestive biscuits
75g butter

Line and grease a 20cm springform cake tin, or grease a 20cm flan dish (see point 2 on p310).

Blitz the biscuits to crumbs in a food processor; or put them in a plastic bag and set about them with a rolling pin. Melt the butter over a gentle heat in a saucepan, and stir in the crumbs.

Tip the buttery crumbs into the cake tin or flan dish, spread them out, and compact them with the back of a spoon. Put the tin/dish into the fridge or freezer, to allow the base to firm up while you make the filling.

Use chocolate digestives. Or plain digestives, with a tbsp of cocoa powder. Or ginger biscuits.

LEMON AND LIME CHEESECAKE

HOW TO MAKE IT

For a 20cm springform tin or flan dish
filled with biscuit base (see p343)
142ml double cream
200g full-fat cream cheese
397g tin condensed milk
Juice and zest of 1 lemon
Juice of 2 limes, zest of 1

Whisk the cream until it thickens, but not until it goes stiff; it should be thick but still fluid. Combine the cheese and the condensed milk, and whisk them until they have the same consistency as the cream. Stir in the cream, along with the juice and zest of the citrus fruits. (You could of course use just lemon, or just lime.)

Pour the mixture over the biscuit base in the tin or dish.

Cover the dish or tin with foil, tented at the top so that it does not stick to the filling, and refrigerate for at least 3 hours. Release from the tin, if using.

CHOCOLATE CHEESECAKE

This cheesecake is based loosely on one in *Classic Cheese Cookery* by Peter Graham (Grub Street). Graham also includes 3 limes, crème de menthe, and mint leaves; his cheesecake contains a hefty 280g chocolate. The disadvantage of my quantity is that the pale brown of the filling is not particularly attractive. But 280g would be a bit much, I think. You could leave out the chocolate altogether, and just have lime juice (and zest), or lemon, or perhaps a combination of the two as in the version above.

HOW TO MAKE IT

For a 20cm springform tin or flan dish
filled with biscuit base (see p343)
3 tsp gelatine, or 3 sheets of leaf gelatine
500g ricotta or cottage cheese (drain the cottage cheese)
397g tin condensed milk (about 300ml)
200ml double cream, whipped until slightly thickened
100g dark chocolate

Put about 4 tbsp cold water into a small saucepan. Sprinkle over the gelatine (or add the sheets), and swirl the water about to cover. Add more water if necessary. Set aside.

In a bowl, stir together the ricotta, condensed milk and cream until thoroughly blended. (You may prefer to use 500ml cream alone, without the condensed milk. In which case, add 60g caster sugar too.)

Place a bowl in a saucepan of gently simmering water so that the base of the bowl does not touch the water. Break up the chocolate, put it in, and stir until melted. Remove from the heat.

Add a few spoonfuls of the cheese and cream mixture to the chocolate – stirring them together will help to release the chocolate from the side of the bowl. Tip this mixture into the bowl of cheese and cream, and blend.

Put the saucepan with the gelatine on to the lowest possible flame, and stir. As soon as the gelatine dissolves and the mixture clarifies, remove it from the heat. (Boiling gelatine disables its setting qualities.) Keep stirring until thoroughly dissolved. Pour the gelatine into the cheese mixture, and blend thoroughly.

Pour the mixture over the biscuit base in the tin or dish. Cover the dish or tin with foil, tented at the top so that it does not stick to the filling, and refrigerate for at least 3 hours. Release from the tin, if using.

NEW YORK CHEESECAKE

A baked cheesecake. Baked cheesecakes tend to be fudgier in texture; but the ricotta or cottage cheese keeps this one light.

Use a 23cm springform tin for this recipe, and make your base with 175g of biscuits and 85g of butter (see p343)
250g ricotta, or cottage cheese (drain the latter)
250g cream cheese
125g caster sugar
2 eggs, separated[1] – SEE HOW YOU DO IT
125ml sour cream
2 tbsp flour
1 tsp vanilla essence
Zest from 1 lemon

Heat the oven to gas mark 4/180°C, and put in a baking sheet.

Blend the cheeses. Beat in the sugar and egg yolks. Whip the cream until it starts to thicken, and stir it into the mixture with the flour, vanilla essence and lemon zest. Beat with a wooden spoon until smooth. Whisk the egg whites until they form soft peaks,[2] and fold them gently into the mixture. Pour the mixture on to the crust in the cake tin. Spread smooth.

Put the tin on to the baking sheet (which aids heat transmission) in the oven, and bake for about an hour, or until the cheesecake is no longer wobbly in the centre.

Allow the cake to cool. Then chill it in the fridge. When it's properly chilled, release it from the tin.

I am not very keen on the sweet, fruity toppings that sometimes come with cheesecakes. Perhaps I wouldn't mind the simple blueberry compote on p331.

WHY YOU DO IT

1 • **Separating the eggs**. See point 5 on p311.

2 • **Whisking the whites**. See point 6 on p311.

KEY LIME PIE

Not a cheesecake, but related to one. This recipe is adapted from Nigella Lawson's Kitchen.

HOW TO MAKE IT

*For a 20cm springform tin or flan dish filled with
    biscuit base (see p343)*
397g tin condensed milk
3 limes – juice and zest
284ml double cream

Pour the condensed milk into a bowl. Stir in the lime juice and zest, and pour in the cream. Whisk. With a hand whisk, this is quite hard work. The mixture does not go stiff, but it does, eventually, thicken.

Pour the mixture over the biscuit base in the tin or dish. Cover the dish with foil, tented at the top so that it does not stick to the filling, and refrigerate for at least 3 hours, or overnight. Release from the tin, if using.

BIBLIOGRAPHY

Top three

For the cooking we do every day, recipes are less useful than is a set of principles, techniques and templates, from which we can create our own recipes. Here are the three books that are the best I know at inspiring that improvisational confidence.

Simple French Food by Richard Olney (Grub Street). 'It ought to be called Incredibly Complicated French Food,' someone once said to me. 'Simple', in this context, does not mean 'basic', but indicates that Olney is dealing with regional cooking rather than with high gastronomy. Actually, there are plenty of recipes here that are far from complicated; but only the most dedicated reader is likely to follow Olney every step of the way in his 'Sauté-type' stew, or in his fish terrine with whipped tomato cream. The real genius of *Simple French Food*, though, is in the prose. Olney, who died in 1999, was an American who lived in France, and he wrote with a patrician fastidiousness reminiscent of another American expat, Henry James. *Simple French Food* is about the principles and, it is not going too far to say, philosophy underlying the preparation of classic dishes. 'By knowing and accepting rules, one frees oneself of rules,' he summarized. Even if you never cook a recipe from this book, you'll find it a revelation.

Real Fast Food by Nigel Slater (Penguin). What Olney does for classic French cuisine, Slater does, in an appropriately chattier style, for the kind of cooking most of us do after a trip to the deli and a raid on the store cupboard. Again, it's not the recipes that make this book stand out, although they are worth more than the cover price, but the emphasis on improvisation. That is the guiding principle of all Slater's books, particularly *Appetite* (Fourth Estate), with its series of recipe templates.

McGee on Food and Cooking by Harold McGee (Hodder & Stoughton). This is an 850-page book on the science, history and culture of the kitchen. Don't be daunted: Harold McGee is a most engaging, companionable guide to these topics. After reading him, you'll know why a sauce thickens, what happens when you make a mayonnaise, how vegetables lose colour, whether flash-frying seals meat. It will put you in a better position to assess the recipes of others; and it will give you confidence to adopt your own techniques, based on the soundest principles.

Some more books I like:

The Pedant in the Kitchen by Julian Barnes (Atlantic)
The Science of Cooking by Peter Barham (Springer) – aimed at schoolchildren, but of interest to anyone
Classic Turkish Cookery by Ghillie and Jonathan Basan (I B Tauris)
The Gastronomy of Italy by Anna del Conte (Pavilion)
Four Seasons Cookery Book by Margaret Costa (Grub Street)
French Provincial Cooking by Elizabeth David (Penguin)
The Perfect ... by Richard Ehrlich (Grub Street)
The River Cottage Meat Book by Hugh Fearnley-Whittingstall (Hodder & Stoughton)
Classic Cheese Cookery by Peter Graham (Grub Street)
Jane Grigson's Vegetable Book (Penguin)
Marcella's Kitchen by Marcella Hazan (Macmillan)
Nose to Tail Eating by Fergus Henderson (Bloomsbury)

How To Cook Better by Shaun Hill (Mitchell Beazley)
Geraldine Holt's Cakes (Prospect Books)
Curry Easy by Madhur Jaffrey (Ebury)
How To Eat by Nigella Lawson (Chatto & Windus)
Keep It Simple by Alastair Little and Richard Whittington
(Conran Octopus – but out of print)
Made in Italy by Giorgio Locatelli (Fourth Estate)
The Rice Book by Sri Owen (Frances Lincoln)
The Food of Italy by Claudia Roden (Vintage)
Sauces by Michel Roux (Quadrille)
How To Cook by Delia Smith (3 volumes, BBC)
The Man Who Ate Everything by Jeffrey Steingarten (Head-
line)
The Cook's Encyclopaedia by Tom Stobart (Grub Street)
Home Cooking by Richard Whittington (Cassell)

ACKNOWLEDGEMENTS

Emma Hayes of the Food Standards Agency answered all my queries informatively, and with patience. Darren Chant of Basingstoke and Deane Borough Council also gave me useful advice on food safety. Claire Domoney of the John Innes Centre in Norwich furthered my understanding of the cooking of legumes, as did Geoffrey Kite at Kew.

I'd like to thank also Cortina Butler, Anne Dolamore, Fiona Hunter, Gwilym Lewis, Marion Regan and Betsy Tobin.

INDEX

acid 271
aioli 42–3
alcohol 34–5
  aligot 178–9
  almond cake 318–19
  Alsace onion tart 98–9
anchovies: anchovy, garlic and chilli
    sauce 123–5
  and broccoli 148
and cauliflower 154
puttanesca sauce 197–8
apples: apple crumble 335–6
  apple sauce 219
apricots 332,338
arrowroot 232
asparagus 139-40
aubergines 140–3
  moussaka 258–9
  Parmigiana di melanzane 143
  ratatouille 173–5
  with pesto 40–1

bacon 265,279
  Alsace onion tart 99
  bacon risotto 107
  in macaroni cheese 128
  quiche Lorraine 95–6
  see also pancetta
bacteria 55,77–8,208
bain-maries 323–4
balloon whisks 28
barbecues 271,272–3,278,296
Barham, Peter 312
Barnes, Julian 8
Basan, Ghillie 140–1
basil: pesto 39–41
batter: clafoutis 337–9
  fish in 294–5
  Yorkshire pudding 225–7
beans: bean soup 70–3
  cassoulet 247–9 dried beans 144–7
  green beans 143–4
béchamel sauce 9–10,46–8,125–9
beef: beef stew 229–31
  boeuf Bourguignon 228,233–4
  carbonnade 228
  cottage pie 260–2
  meatballs 262–4
  mince 255–6
  oxtail stew 238–9
  ragù alla Bolognese 256–8,262
  roast beef 224–5
  steak 201,234,269–70,273–4
beer batter 295

bicarbonate of soda 167
blackberries 332,336
blackcurrants 332,334
  blackcurrant crumble 336
Blanc, Raymond 41,232,337,338
blanquette 228,252–3
blenders 27,66
blind baking pastry 95
Blumenthal, Heston 179,186,326
bobby beans 143–4
boeuf Bourguignon 228,233–4
Bolognese sauce 125–7,256–8,262
bones, stock 56
bouillabaisse 302,303
Boxer, Arabella 339
braises 228
brandy, flaming 35
bread: anchovy, garlic and chilli sauce
    123–5
  bread sauce 49
  cheese on toast 90
  croutons 232–3
  fried cheese sandwich 91
  soup garnish 62
bread and butter pudding, panettone
    324–5
brine 32
  brined and roasted pork belly 218–19
broccoli 147–9
Brussel tops 152
butter: clarified butter 190
  creaming 310–11
  hollandaise sauce 44–6
  and oil 293
  in scrambled eggs 84
  splitting 269

cabbage 149–50
cake tins 13,310
cakes 308–20
  almond cake 318–19
  chocolate cake 316–17
  Genoese sponge 316
  lemon polenta cake 319–20
  sponge cakes 310–19
  Victoria sponge 314–15
  without flour 317–20
calabrese 148
cannellini beans 144
caper sauce 221–2
carbonnade of beef 228
carbonara sauce 118–21
carrots 152–3
  carrot and coriander soup 64–5